SmartMaths
Standards 4 and 5

Karen Morrison
Lisa Greenstein

Consultant editor:
Penelope Furlonge

The answers, glossary and curriculum mapping can be found at
http://www.hachettelearning.com/mathematics/smartmaths-standards-4-and-5

The Publishers would like to thank the following for permission to reproduce copyright material.

Photo credits

p. 2 *cc*, **p. 18** *tr* © Andrey Zhuravlev/stock.adobe.com; **p. 4** *tr* © Olga Moonlight/stock.adobe.com; **p. 10** *tr* © PA Images/Alamy Stock Photo; **p. 11** *cl* © JackF/stock.adobe.com; **p. 11** *cc* © MichiruKayo/stock.adobe.com; **p. 11** *cr* © Tarik GOK/stock.adobe.com; **p. 20** *tl* © Joe/stock.adobe.com; **p. 20** *tr* © John de la Bastide/stock.adobe.com; **p. 21** *cr* © Hachette; **p. 24** *tr* © Volha Drabovich/stock.adobe.com; **p. 25** *cc* © Roman/stock.adobe.com; **p. 32** *br* © Mazur Travel/stock.adobe.com; **p. 42** *tr* © bohbeh/stock.adobe.com; **p. 42** *cr* © akhenatonimages/stock.adobe.com; **p. 51** *tr* © New Africa/stock.adobe.com; **p. 55** *cl* © Natali/stock.adobe.com; **p. 57** *cr*, **p. 60** *bc* © sirikornt/stock.adobe.com; **p. 60** *bl* © mything/stock.adobe.com; **p. 60** *br* © reshoot/stock.adobe.com; **p. 61** *cr* © Tanya/stock.adobe.com; **p. 62** *tr* © Photobeps/stock.adobe.com; **p. 65** *tc* © yusuf/stock.adobe.com; **p. 65** *tr* © Christian/stock.adobe.com; **p. 65** *cr* © mdbildes/stock.adobe.com; **p. 65** *cl* © moodboard/stock.adobe.com; **p. 65** *cc* © ChayTee/stock.adobe.com; **p. 65** *cr* © Golden Sikorka/stock.adobe.com; **p. 76** *tr* © Jamie Squire/Staff/Getty Images; **p. 78** *tc* © Valerii Evlakhov/stock.adobe.com; **p. 78** *tr* © Aratchaporn/stock.adobe.com; **p. 78** *cc* © New Africa/stock.adobe.com; **p. 78** *cr*, **p. 146** *br* © Winai Tepsuttinun/stock.adobe.com; **p. 78** *cc* © VitaL/stock.adobe.com; **p. 78** *cr* © Gayan/stock.adobe.com; **p. 78** *cl* © Juulijs/stock.adobe.com; **p. 78** *cc* © Gudman/stock.adobe.com; **p. 81** *bl* © daboost/stock.adobe.com; **p. 83** *cc* © Elena/stock.adobe.com; **p. 83** *cr* © New Africa/stock.adobe.com; **p. 87** *cl* © Nikolai Sorokin/stock.adobe.com; **p. 90** *br* © Anatoliy Karlyuk/stock.adobe.com; **p. 90** *cr* © sablin/stock.adobe.com; **p. 93** *br* © muratart/stock.adobe.com; **p. 94** *tr* © Svetlana/stock.adobe.com; **p. 96** *br* © GLandStudio/stock.adobe.com; **p. 97** *cr* © SockaGPhoto/stock.adobe.com; **p. 107** *tr* © Iliya Mitskavets/stock.adobe.com; **p. 113** *cr* © NIKCOA/stock.adobe.com; **p. 119** *tr* © Sergey Ryzhov/stock.adobe.com; **p. 124** *tc* © Udayakumar/stock.adobe.com; **p. 125** *cr* © Christina Patricya/stock.adobe.com; **p. 130** *cr* © Goldman/stock.adobe.com; **p. 141** *tr* © jedi-master/stock.adobe.com; **p. 143** *cr* © Ekahardiwito/stock.adobe.com; **p. 143** *br* © sosiukin/stock.adobe.com; **p. 144** *cr* © Bbrightiee/Shutterstock.com; **p. 144** *cl* © udaya fire/Shutterstock.com; **p. 144** *cr* © udaya fire/Shutterstock.com; **p. 146** *bc* © injenerker/stock.adobe.com; **p. 149** © udaya fire/Shutterstock.com; **p. 153** *br* © SockaGPhoto/stock.adobe.com; **p. 157** *tr* © John de la Bastide/Shutterstock.com; **p. 158** *tr* © RayBond/stock.adobe.com; **p. 160** *tr* © Genzi/stock.adobe.com; **p. 160** *cr* © Australian Associated Press/Alamy Stock Photo; **p. 166** *tr* © cgdeaw/stock.adobe.com; **p. 169** *tr* © Marina Marr/stock.adobe.com; **p. 173** *tl* © dana_zurki/stock.adobe.com; **p. 179** *tr* © dana_zurki/stock.adobe.com; **p. 183** *br* © vvoe/stock.adobe.com; **p. 186** *tr* © tasty_cat/stock.adobe.com; **p. 197** *cl* © Dejan Jovanovic/stock.adobe.com; **p. 198** *br* © Eric Isselée/stock.adobe.com; **p. 207** *bc* © pixelrobot/stock.adobe.com; **p. 209** *tc* © Andreas Liem/Shutterstock.com; **p. 210** *cr* © New Africa/stock.adobe.com; **p. 218** *cc* © Fleur Paper Co/Shutterstock.com.

Although every effort has been made to ensure that website addresses are correct at time of going to press, Hachette Learning cannot be held responsible for the content of any website mentioned in this book. It is sometimes possible to find a relocated web page by typing in the address of the home page for a website in the URL window of your browser.

Hachette UK's policy is to use papers that are natural, renewable and recyclable products and made from wood grown in well-managed forests and other controlled sources. The logging and manufacturing processes are expected to conform to the environmental regulations of the country of origin.

To order, please visit www.HachetteLearning.com or contact Customer Service at education@hachette.co.uk / +44 (0)1235 827827.

ISBN: 9781036012106

@ Lisa Greenstein & Karen Morrison 2025

First published in 2025 by Hachette Learning,
An Hachette UK Company
Carmelite House
50 Victoria Embankment
London EC4Y 0DZ

www.HachetteLearning.com

The authorised representative in the EEA is Hachette Ireland, 8 Castlecourt Centre, Dublin 15, D15 XTP3, Ireland (email: info@hbgi.ie)

Impression number   10 9 8 7 6 5 4 3 2 1

Year   2029 2028 2027 2026 2025

All rights reserved. Apart from any use permitted under UK copyright law, no part of this publication may be reproduced or transmitted in any form or by any means, electronic or mechanical, including photocopying and recording, or held within any information storage and retrieval system, without permission in writing from the publisher or under licence from the Copyright Licensing Agency Limited. Further details of such licences (for reprographic reproduction) may be obtained from the Copyright Licensing Agency Limited, www.cla.co.uk

Cover illustration by Arminda Bailey

Illustrations by Stéphan Theron & Vian Oelofsen

Typeset in FS Albert 12 in 14 pt by IO Publishing

Printed and bound in Great Britain by Bell & Bain Ltd, Glasgow

A catalogue record for this title is available from the British Library.

Contents

Section 1 Number
- Chapter 1 — Numbers, place value and rounding
- Chapter 2 — Number patterns and number relationships
- Chapter 3 — Number operations

Section 2 Geometry
- Chapter 4 — Angles
- Chapter 5 — Solids and plane shapes

Section 3 Number
- Chapter 6 — Fractions
- Chapter 7 — Decimals

Section 4 Measurement
- Chapter 8 — Measuring length, mass and time

Section 5 Statistics
- Chapter 9 — Handling data

Section 6 Measurement
- Chapter 10 — Perimeter and area
- Chapter 11 — Capacity and volume

Section 7 Problem solving
- Chapter 12 — Solving real-life problems

Section 8 Measurement
- Chapter 13 — Mass and weight

Section 9 Number
- Chapter 14 — Fractions 2
- Chapter 15 — Decimals 2

Section 10 Geometry
- Chapter 16 — Solids and plane shapes 2

Section 11 Measurement
- Chapter 17 — Perimeter and area 2

Section 12 Number
- Chapter 18 — Percentages

Section 13 Statistics
- Chapter 19 — Handling data 2

Section 14 Measurement
- Chapter 20 — Volume

Section 15 Exam preparation
- Chapter 21 — Prepare for the Secondary Entrance Assessment (SEA)

The answers, glossary and curriculum mapping can be found at www.hachettelearning.com/mathematics/smartmaths-standards-4-and-5

How to use this book

This Student's Book meets the content, skills, dispositions, outcomes and elaborations specified for Standards 4 and 5 in the primary curriculum for Trinidad and Tobago. Alongside the full coverage of the mathematics curriculum, the book is also designed to address the Assessment Framework for the Secondary Entrance Assessment (SEA).

The mathematics curriculum is divided into four broad strands: Number, Geometry, Measurement and Statistics. Each section in the book corresponds to a strand in the curriculum, and chapters are colour-coded to show which strand they belong to. The strands have been spiralled in a logical order so that students encounter concepts from all four strands over a two-year period.

More information on curriculum coverage can be found in the 'Curriculum mapping' document on our website. Answers to all questions in this book can also be found there.

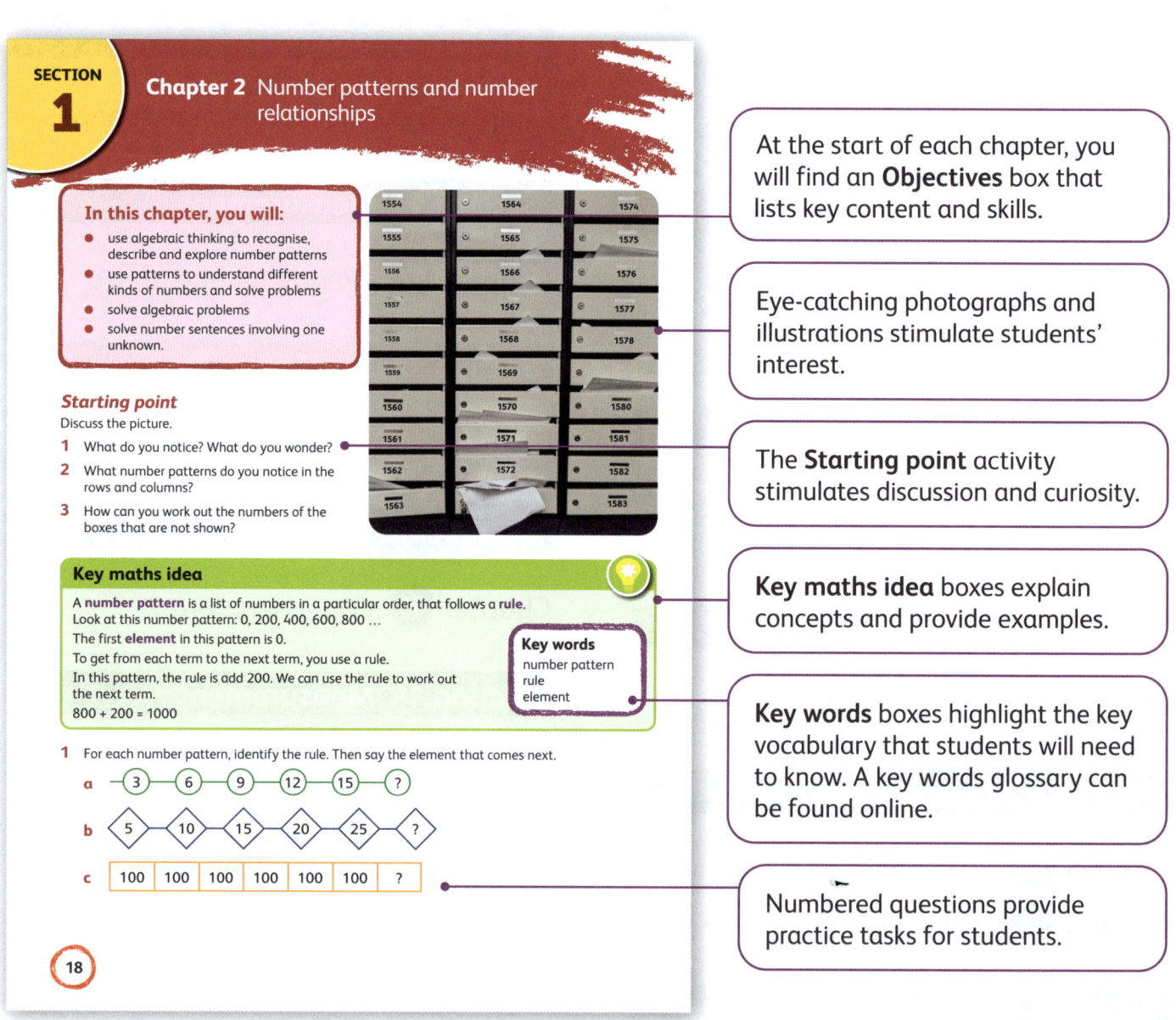

At the start of each chapter, you will find an **Objectives** box that lists key content and skills.

Eye-catching photographs and illustrations stimulate students' interest.

The **Starting point** activity stimulates discussion and curiosity.

Key maths idea boxes explain concepts and provide examples.

Key words boxes highlight the key vocabulary that students will need to know. A key words glossary can be found online.

Numbered questions provide practice tasks for students.

How to use this book

As students work through the chapters, they will find a range of features, including:
- **Talking maths:** This feature supports vocabulary and communication skills.
- **Real-life maths:** This feature links mathematical concepts to real-world applications.
- **Mental maths:** This feature includes application of mental maths skills and questions.
- **Full STEAM ahead:** This feature provides activities that integrate mathematics with science, technology, engineering and the arts.
- **Problem solving:** This feature gives students a chance to apply their mathematical skills to problem-solving scenarios using the strategies they have learnt.
- **Hints:** This feature supplies additional background information, reminders or links to concepts that have appeared elsewhere.
- **Calculator:** This symbol appears next to questions where students are encouraged to use a calculator.
- **Extension:** This feature engages with material that goes beyond the curriculum. It presents additional questions for students who need an extra stretch or challenge.
- **What did you learn?:** This feature appears towards the end of each chapter. It assesses how confident students are feeling about the material they have just covered.

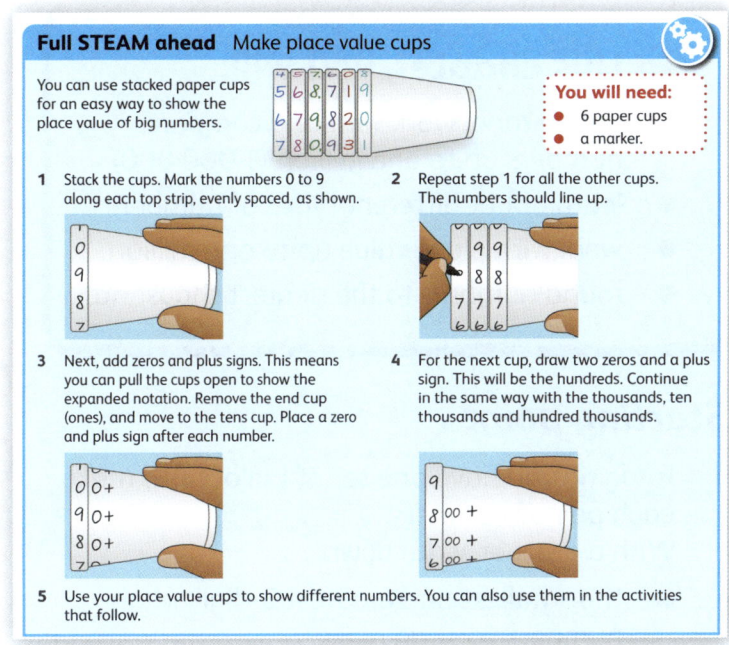

Each chapter ends with a **Review** page, which comprises:
- **Key terms and concepts:** vocabulary questions that check students understand the main terms and concepts presented in the chapter.
- **Quick check:** short questions that revise the mathematics covered through the chapter.
- **Challenge and investigate:** long questions or activities that can extend beyond the main curriculum for students who need additional extra challenges.

At the end of this book, there is a full chapter devoted to preparation for the SEA. This chapter begins with tips and exam skills to help students understand exactly what they will need to do in the exam, and how to prepare effectively. The chapter then provides a range of practice questions to allow students to become comfortable with the format and expectations of the exam questions.

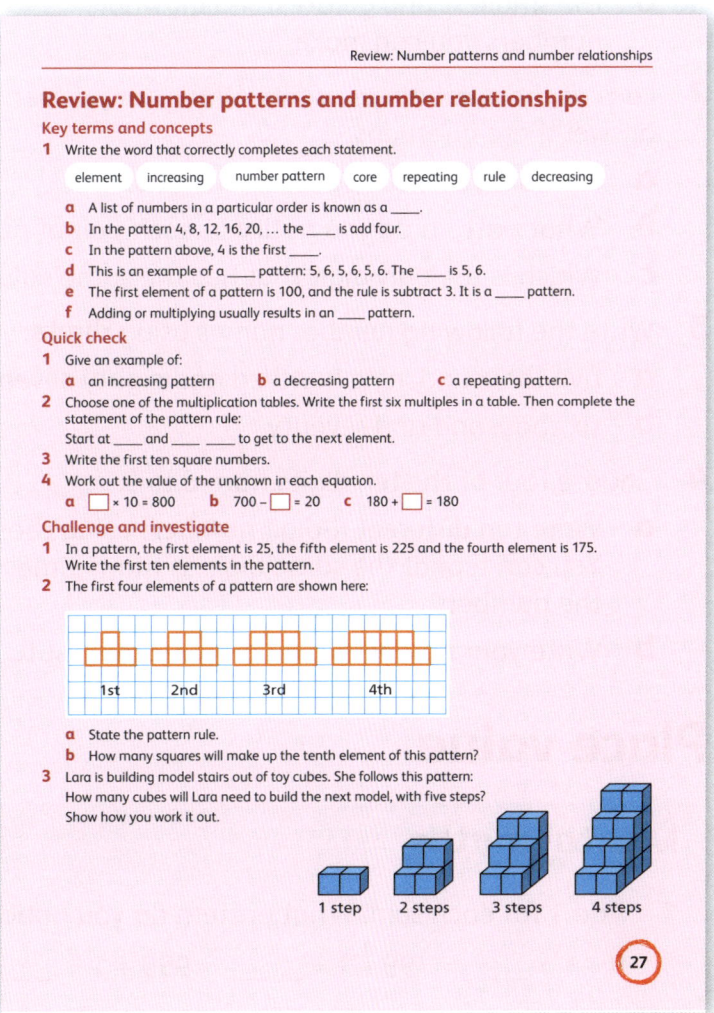

SECTION 1

Chapter 1 Numbers, place value and rounding

In this chapter, you will:
- understand, work with and compare numbers up to one million (1 000 000)
- learn about different types of numbers
- work with place value up to one million
- round numbers to the nearest thousand.

Starting point

1. Imagine you have one set of balloons to make each pair of numbers.
   With a partner, write down:
   a the smallest and greatest 3-digit whole numbers you can make
   b the smallest and greatest 4-digit whole numbers you can make.

2. Look at the numbers you wrote. With your partner, answer these questions.
   a How did you work out your answers?
   b What pattern do you notice in each pair of numbers?
   c Which is the only digit that has the same value no matter which place it is in? Why?

3. Write the following number names using **digits**:
   a five thousand, nine hundred and eighty-seven
   b six thousand and seventy

4. Choose four of the ten digits from 0 to 9.
   a Write ten different 4-digit numbers using those four digits. You can repeat the same digit as many times as you like in the numbers.
   b Write your numbers in order from smallest to greatest.

Key word
digit

Place value

Mental maths

1. Add 1 for each sum. What pattern do you notice?

   9 + 1 = ____   99 + 1 = ____   999 + 1 = ____   9999 + 1 = ____

4

Place value

Key maths idea

Our number system uses ten digits: 0, 1, 2, 3, 4, 5, 6, 7, 8 and 9.
We use **place value** to make greater numbers.

Key words
place value
thousand
ten thousand
one hundred thousand

Example

This seed packet contains 100 seeds.

H	T	O
1	0	0

This box contains 10 seed packets.

10 groups of 100 make one **thousand**:

Th	H	T	O
1	0	0	0

A store sells seeds to 10 customers.
Each customer buys 10 boxes.
10 groups of 1000 make **ten thousand**:

T Th	Th	H	T	O
1	0	0	0	0

The store sells 10 boxes every day for 10 days.
Each box has 10 000 seeds.
10 groups of 10 000 make 100 000
(**one hundred thousand**).

H Th	T Th	Th	H	T	O
1	0	0	0	0	0

---

1 Write the total value. The first one has been done for you.
   a 10 tens = <u>100</u>   **b** 10 hundreds =   **c** 10 thousands =   **d** 10 ten thousands =

2 Write the number of seeds you have if you buy:
   a one box of 1000 seeds and one more packet of 100
   b three packets of 100 and two boxes of 1000 seeds.

3 If the store sells ten thousand seeds per day, how many do they sell in:
   a 2 days   **b** 5 days   **c** 14 days?

4 Write these number names in order from greatest to smallest.

    one thousand   seventeen thousand

    twenty-five thousand   three hundred thousand

5 Write each set of numbers in order from smallest to greatest.
   a 10 000   1   100   100 000   1000
   b 5000   300 000   30 000   300   500   3500

6 Draw a place value table to show each number name in digits.
   a thirty-five thousand, four hundred and fifty-two
   b five hundred and six thousand, nine hundred and eighty-nine

7 Write the place value of the 5 in each of the following numbers.
   a 135 467   **b** 659 301   **c** 428 560

Section 1 **Number** Chapter 1 Numbers, place value and rounding

Digits, expanded notation and number names

> **Key maths idea**
>
> Think about the number 7542. Count on in thousands:
>
> 7542, 8542, 9542, 10 542, 11 542, 12 542, 13 542
>
> You cannot have more than nine thousands in the thousands column. When you reach ten thousand, you need another column to the left in the **place value chart**. This is the ten thousands place. Look at this number: 13 542
>
> | T Th | Th | H | T | O |
> |---|---|---|---|---|
> | 1 | 3 | 5 | 4 | 2 |
>
> **Expanded notation**
>
> 13 542 = (1 × 10 000) + (3 × 1000) + (5 × 100) + (4 × 10) + (2 × 1)
>
> Or: 13 542 = 10 000 + 3000 + 500 + 40 + 2
>
> Number name: thirteen thousand, five hundred and forty-two
>
> When you reach one **hundred thousand**, you need another column to the left. This is the hundred thousands place. Look at this number: 163 242
>
> | H Th | T Th | Th | H | T | O |
> |---|---|---|---|---|---|
> | 1 | 6 | 3 | 2 | 4 | 2 |
>
> Say: 'one hundred and sixty-three thousand, two hundred and forty-two'.

Key words
place value chart
hundred thousand
expanded notation

Hint
Sometimes a question asks you to express a number using **expanded notation**. If the question does not give an example, you may choose whether or not to use brackets.

1 Express each number on a place value table and in expanded notation.
   a 45 213   **b** 142 019   **c** 18 055   **d** 741 240

2 Each number is given in expanded notation. Write the number name.
   a (4 × 10 000) + (0 × 1000) + (5 × 100) + (1 × 10) + (3 × 1)
   b (9 × 10 000) + (2 × 1000) + (0 × 100) + (7 × 10) + (8 × 1)
   c (8 × 10 000) + (7 × 1000) + (4 × 100) + (3 × 10) + (0 × 1)
   d (7 × 100 000) + (1 × 10 000) + (9 × 1000) + (0 × 100) + (3 × 10) + (7 × 1)
   e (5 × 100 000) + (0 × 10 000) + (6 × 1000) + (4 × 100) + (7 × 10) + (6 × 1)

3 Say each number name. Choose three of them to write in expanded notation.
   a 234 000   **b** 300 000   **c** 576 234   **d** 703 000
   e 499 624   **f** 100 095   **g** 897 280   **h** 385 040

4 Write each number using digits.
   a three hundred and ninety-one thousand, four hundred and seventy-three
   b eight hundred thousand, three hundred and seven
   c nine hundred and three thousand, five hundred and eighteen

5 Write the value of the 4 in each number.
   a 456 009   **b** 376 214   **c** 896 437   **d** 205 842

Digits, expanded notation and number names

Full STEAM ahead   Make place value cups

You can use stacked paper cups for an easy way to show the place value of big numbers.

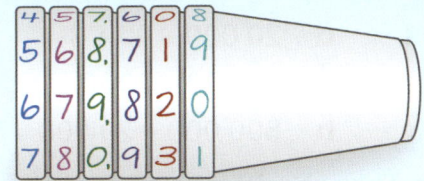

You will need:
- 6 paper cups
- a marker.

1. Stack the cups. Mark the numbers 0 to 9 along each top strip, evenly spaced, as shown.

2. Repeat step 1 for all the other cups. The numbers should line up.

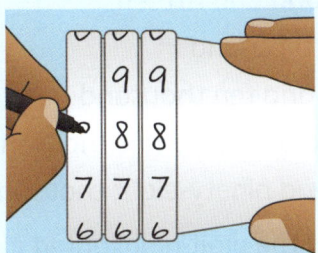

3. Next, add zeros and plus signs. This means you can pull the cups open to show the expanded notation. Remove the end cup (ones), and move to the tens cup. Place a zero and plus sign after each number.

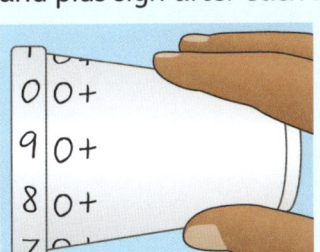

4. For the next cup, draw two zeros and a plus sign. This will be the hundreds. Continue in the same way with the thousands, ten thousands and hundred thousands.

5. Use your place value cups to show different numbers. You can also use them in the activities that follow.

Key maths idea

We can use a place value chart to organise numbers by value, in other words, to arrange them from smallest to greatest. These cars have prices in hundreds of thousands. Here are the prices in a place value table. Say each number aloud.

H Th	T Th	Th	H	T	O
3	5	3	0	0	0
2	2	5	0	9	9
1	5	9	0	8	5

$353,000

$225,099

$159,085

To see which is the highest price, compare the digits with the greatest place value. The prices all have six digits. Their greatest place value is hundred thousands. $353 000 has 3 in the hundred thousands place. This is the highest price.

Section 1 Number Chapter 1 Numbers, place value and rounding

1. Write the numeral for each expanded notation. The first one has been done for you.
   - a 3000 + 400 + 50 + 8 = 3458
   - b 6000 + 20 + 9
   - c 10 000 + 2000 + 700 + 80 + 1
   - d 20 000 + 40 + 3
   - e 90 000 + 4000 + 500 + 2
   - f 70 000 + 9000 + 30
   - g 300 000 + 9000 + 50 + 4
   - h 800 000 + 20 000 + 800 + 80

   Hint
   If you have made place value cups, use them to make the numbers.

2. Write these numbers using digits. Then check a partner's work.
   - a ten thousand and twenty
   - b eleven thousand and fifteen
   - c one hundred thousand
   - d two hundred and ten thousand
   - e three hundred and forty-five thousand
   - f six hundred and ninety-eight thousand and fifty-seven

3. Write the value of the digit highlighted in colour.
   - a 49 1**3**0
   - b **5**9 428
   - c 19 **3**02
   - d 4**6** 616
   - e 740 49**8**
   - f 3**9**9 999
   - g **7**84 033
   - h 271 7**4**7

4. Write the number name for each numeral.
   - a 10 487
   - b 14 051
   - c 42 075
   - d 73 958
   - e 104 039
   - f 284 803
   - g 972 099
   - h 703 001

5. Look at this example.
   52 308 = 5 ten thousands, 2 thousands, 3 hundreds, 0 tens, 8 ones
   Write each number in the expanded form shown.
   - a 18 377
   - b 26 439
   - c 71 898
   - d 30 856
   - e 478 207
   - f 500 214
   - g 375 753
   - h 101 221

6. Make up three different six-digit numbers. Write them as number names. Check a partner's work.

7. Write each number in expanded form.
   Example: 451 298 = 400 000 + 50 000 + 1000 + 200 + 90 + 8
   - a 829 407
   - b 482 371
   - c 314 099
   - d 840 570

8. Write each number in expanded form.
   Example: 60 392 = (6 × 10 000) + (3 × 100) + (9 × 10) + (2 × 1)
   - a 683 914
   - b 932 010
   - c 209 957

Up to one million

Talking maths

What do these expressions mean? Discuss your ideas with a partner.

- That is the million-dollar question.
- Thanks a million!
- You are one in a million.

Key maths idea

999 999 + 1 = 1 000 000   one **million**

You can see from the zeros that one million is one thousand thousands. You can also think of it as ten groups of 100 000.

M	H Th	T Th	T	H	T	O
1	0	0	0	0	0	0

Key words
million
place value mat

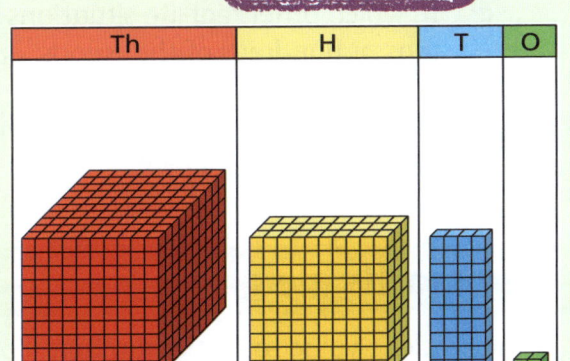

A **place value mat** can help you to see numbers up to thousands. Instead of writing a number in each column, use objects or shapes.

For numbers up to thousands, you can use blocks to show ones, sticks of ten blocks to show tens, squares or bundles of ten to show hundreds and cubes to show thousands. For bigger numbers, it can help to use different shapes.

1. Write the number names of these numbers.
   - a 901 999
   - b 902 000
   - c 999 000
   - d 999 999
   - e 1 000 000
   - f 990 590

Hint
If you do not have place value mats in your classroom, you can make your own.

2. Write the ten numbers that come after 999 990.

3. Lee used objects on a place value mat to represent the number 124 523. Look at what he did.
   - a How many of the heart shapes does the pink circle represent?
   - b How many of the brown cubes does each green star represent?

4. Use a place value table to show:
   - a 540 802
   - b 103 876
   - c one million

Extension

5. Use your place value table to show:
   - a half a million
   - b 3000 less than 248 478
   - c 15 000 more than 143 900

Section 1 **Number** Chapter 1 Numbers, place value and rounding

Rounding to the nearest thousand

Talking maths

1 Read these statements.
   - 'The Hasely Crawford Stadium can hold **about** 27 000 people.'
   - 'In 2023, the Caribbean received **approximately** 3 million cruise visits.'
   - 'There were **around** 140 000 children of school age in Trinidad and Tobago in 2020.'

2 Discuss.
   a Which word in each sentence tells you the numbers are not exact?
   b Why do you think we need numbers that are not exact?
   c In which other real-life situations would you use numbers in the tens or hundreds of thousands, or millions?

3 Two students both **round** the number 4193. Sharon writes 4190. Sasha writes 4200. The teacher says both are correct. Why did they get different answers?

Key words
about
approximately
around
round

Key maths idea

You use place value to round numbers. Follow these steps.

1 Decide which place you are rounding to. For example, if the question says 'round to the nearest hundred', look at the digit in the hundreds place.
2 Look at the digit to the right of the rounding place.
   - If the digit to the right is 0, 1, 2, 3 or 4, the digit in the rounding place stays the same. This is called 'rounding down'.
   - If the digit to the right is 5, 6, 7, 8 or 9, round up by 1.
3 Change all the digits to the right of the rounding place to 0.

Here are two examples:

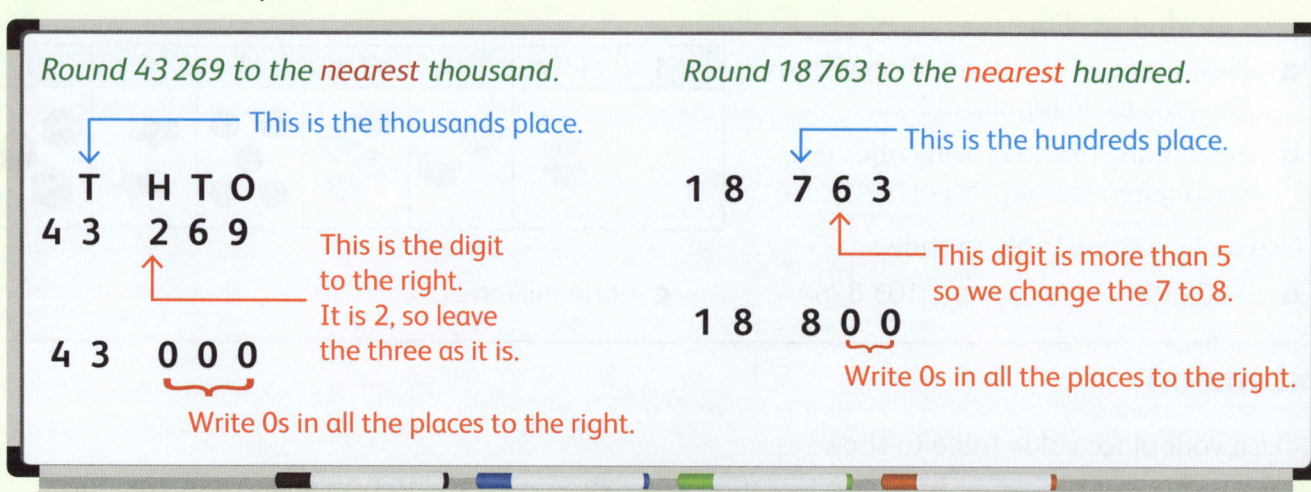

10

Rounding to the nearest thousand

Work in pairs or in your book.

1. Round each number to the nearest hundred.
   - a 2345
   - b 3470
   - c 1399
   - d 6238
   - e 10 260
   - f 32 309
   - g 74 861
   - h 89 055
   - i 411 287
   - j 567 648
   - k 200 962
   - l 791 126

2. Round each number in question 1 to the nearest thousand.

3. Decide which of these numbers will give 6000 when they are rounded to the nearest thousand.

   5098   5890   5499   6023   6500   5500   7023   6499

4. True or false? Give a reason.
   - a 8731 rounds to 9000
   - b 12 458 rounds to 12 500
   - c 112 987 rounds to 114 000

Extension

These questions use numbers that are greater than one million. Try them for a challenge!

5. Look at the populations of these three Central American countries for 2024.

Costa Rica — 5 246 714
Panama — 4 527 961
Belize — 416 656

Round these three populations to the nearest thousand, and use the rounded numbers to estimate the combined population of these three countries.

6. A Caribbean news website reported that The Bahamas had 9 654 000 visitors in 2023.
   - a Do you think this was the exact number of tourists? Give a reason.
   - b Between January and May 2024, the number of cruise visitors to The Bahamas was 3 928 529. Round this to the nearest thousand.

Problem solving

1. A number is rounded to the nearest thousand to get 416 000.
   - a What is the smallest number that could have been rounded to get this answer?
   - b What is the greatest possible number that could have been rounded to get this answer?

2. Can you find this number?
   I am a 4-digit number. I am odd. The digit in the ones place is one more than the digit in the thousands. The digit in the tens place is two less than the digit in the ones. The digit in the hundreds place is the smallest even number. The sum of my digits is 14.
   What number am I?

Section 1 Number Chapter 1 Numbers, place value and rounding

Comparing and ordering numbers

Key maths idea

We use the <, > and = signs to compare numbers.

- \> greater than
- < less than
- = is equal to

Example 1

Which is greater: 4500 or 12 400?
You can use place value to compare the numbers.
You can also use expanded notation:
4500 = 4000 + 500 + 0 + 0
12 400 = 10 000 + 2000 + 400 + 0 + 0

H Th	T Th	Th	H	T	O
		4	5	0	0
	1	2	4	0	0

You can see that 12 400 has 1 in the ten thousands place. 4500 only goes up to thousands.
12 400 > 4500

Example 2

Copy and complete. Fill in the blank with the >, < or = signs.
438 267 ____ 430 999

If two numbers have the same number of digits, start by comparing the values in the highest place value.
Both the numbers have 4 in the hundred thousands place and 3 in the ten thousands place.
43**8** 267 has 8 in the thousands place. 43**0** 999 has 0 in the thousands place.
438 267 > 430 999

Work in pairs or in your book.

1. True or false? Give a reason if it is false.
   - **a** 921 001 > one million
   - **b** half a million = 500 000
   - **c** 582 105 > 977 088
   - **d** 523 599 < 91 599
   - **e** 35 089 < 301 089
   - **f** 728 043 < 1 000 000

2. Write the smallest possible whole number that could complete each statement.
   - **a** 498 999 < ____
   - **b** 745 312 < ____
   - **c** 299 999 < ____

3. Write the greatest possible whole number that could complete each statement.
   - **a** 578 011 > ____
   - **b** 1 000 000 > ____
   - **c** 2 hundred thousand > ____

Hint
< after a number means less than.
\> after a number means greater than.
Remember, these signs always 'point' to the smaller number.

Talking maths

Tell your partner how you worked out the answers to questions **2** and **3**.

4 What is the missing number on each number line?

a   714 800   ☐   715 000   715 100

b   850 000   900 000   ☐   1 million

c   125 000   145 000   ☐   185 000

5 Write each set of numbers in ascending order (from smallest to greatest).
   a   328 938   497 637   1 million   4960   four hundred thousand
   b   sixteen thousand   900 000   25 899   198 000   189 399

6 Write each set of numbers in descending order (from greatest to smallest).
   a   734 721   300 000   37 000   307 007   797 733
   b   488 481   804 408   840 004   480 008   840 000

7 What is the number that comes between each pair of numbers?
   a   247 398 ____ 247 400   b   999 998 ____ 1 000 000
   c   163 038 ____ 163 040   d   783 909 ____ 783 911

8 Identify the incorrect number in each number pattern.
   a   314 578   314 577   314 580   314 581
   b   857 350   857 352   857 355   857 356
   c   798 200   798 300   798 350   798 500
   d   910 550   910 560   910 565   910 580

Multiples and factors

Key maths idea

This number line shows what happens if we count in fives:

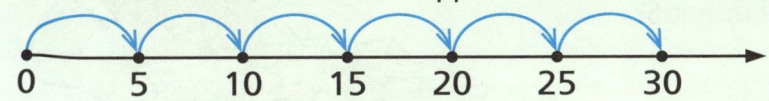

We can count: 0, 5, 10, 15, 20, 25, 30

Each number in a **skip counting** pattern from 0 is a **multiple**.

These are multiples of 5. What are the next three multiples of 5?

The **factors** of a whole number are all the whole numbers which divide into it exactly.

5 is **a prime number**. A prime number has only two factors that can divide into it: the number itself and 1.

10 can be divided by 1, 2, 5 and 10.   Factors of 10: 1, 2, 5 and 10
15 can be divided by 1, 3, 5 and 15.   Factors of 15: 1, 3, 5 and 15
10 and 15 are **composite numbers**. They have other factors besides 1 and themselves.

Key words
skip counting
multiple
factor
product
prime number
composite number

Section 1 **Number** Chapter 1 Numbers, place value and rounding

> **Mental maths**
>
> Work with a partner.
>
> 1 Say the first ten multiples of:
>   a 2   b 3   c 4   d 10
>
> 2 Write out all the factors of:
>   a 16   b 28   c 36
>
> 3 a What is the only prime number that is even?
>   b Why are all the other prime numbers odd?
>
> 4 Keisha says it is impossible for a prime number to end in a zero. Do you agree? Give a reason for your answer.

1 Work out whether each of these numbers is prime or composite.
   a 21   b 31   c 87   d 49   e 99

2 We can use **arrays** to model composite numbers. An array is an arrangement of shapes or objects in columns and rows.
   a Look at this array. Write the multiplication and division facts that it shows.
   b Suggest a different arrangement that would show the same total, but a different pair of factors.

Key word
array

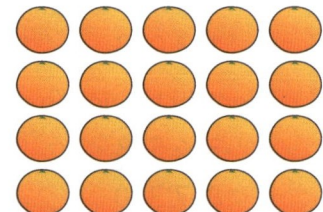

Problem solving

1 Jaden has baked 24 cookies. He wants to share them into boxes with equal numbers of cookies. He has thought of two ways to do it.

   I can make 1 big box with 24 cookies; or 2 boxes, each with 12.

   a If he makes 3 boxes, how many would there be in each box?
   b Why is it possible to make 6 boxes but not 5?
   c What other ways could he do it?

7 is a prime number. Its array would only have one row of seven.

I disagree! It could have seven rows of one.

How many arrays can you make with a prime number? Why?

Factor trees

Key maths idea

Let us look again at the question about Jaden's cookies.

Sadia uses a **factor tree** to work out the answer. A factor tree expresses a composite number as a product of its prime factors. This is called **prime factorisation**.

Key words

factor tree

prime factorisation

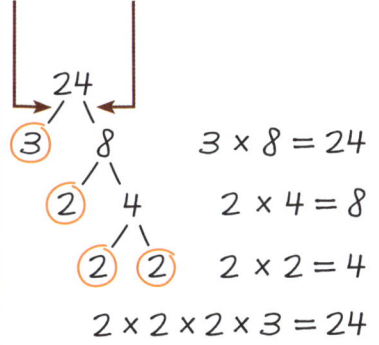

Sadia drew two branches under the number 24, leading to a pair of factors. She used 3 and 8 as the factors because 3 × 8 = 24

24
3
8
3 × 8 = 24
2
4
2 × 4 = 8
2
2
2 × 2 = 4
2 × 2 × 2 × 3 = 24

She chose 3 and 8 as factors because 3 is a prime number. She has circled 3 to show that it is a prime number.

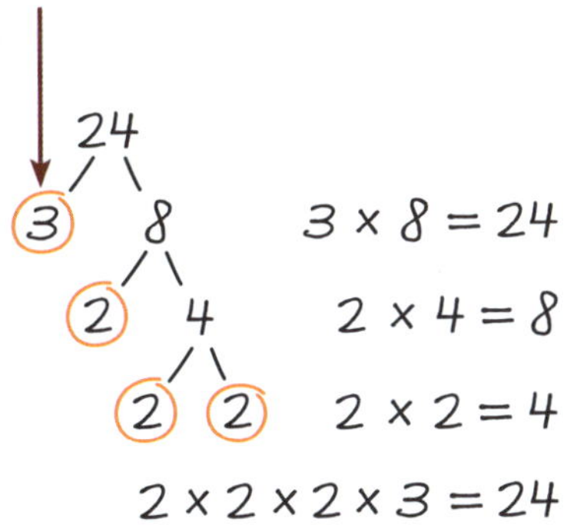

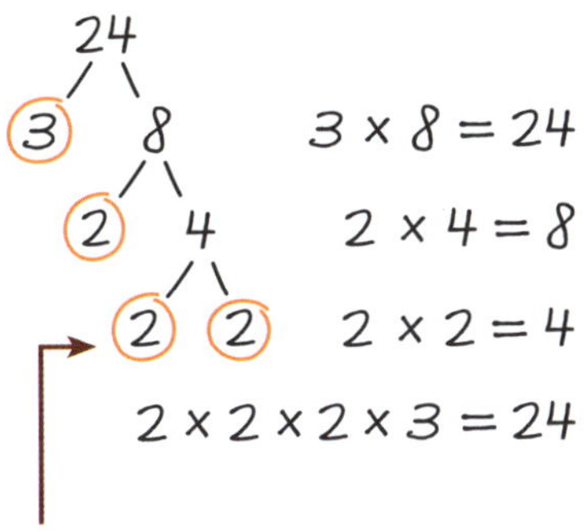

24
3
8
3 × 8 = 24
2
4
2 × 4 = 8
2
2
2 × 2 = 4
2 × 2 × 2 × 3 = 24

24
3
8
3 × 8 = 24
2
4
2 × 4 = 8
2
2
2 × 2 = 4
2 × 2 × 2 × 3 = 24

Sadia keeps factorising until all the factors are prime numbers. She cannot factorise the numbers any further to get a pair of smaller prime numbers.

Sadia can use the factor tree to write 24 as a product of its prime factors. She writes down all the numbers she has circled:

3 × 2 × 2 × 2 = 24

1 Why are some of the numbers not circled in Sadia's factor tree?

2 Six is also a factor of 24, but Sadia did not write 6. Explain why.

3 Copy and complete these factor trees.

a

21

__×__ = 21

b

26

__×__ = 26

c

35

__×__ = 35

d

50

e

45

f

60

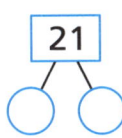

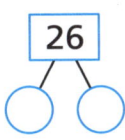

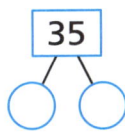

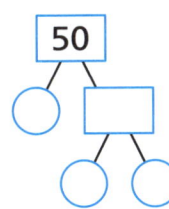

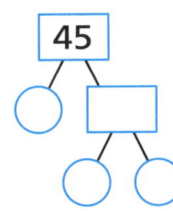

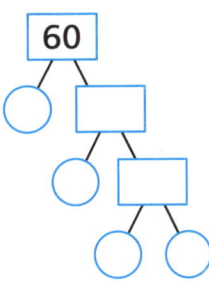

Section 1 Number Chapter 1 Numbers, place value and rounding

Full STEAM ahead   Play the factorisation dice game

Number of players: 2 to 4

Aim: To finish with the highest score

How to play: There are four categories of number in this game. You are trying to get a high score for each of the categories. The categories are:

- an even number
- an odd number
- a multiple of 10
- a multiple of 100.

When it is your turn, roll all four dice.

After the first roll, you may choose to pick up and roll any or all of the dice again. After the second roll, write the four numbers you rolled, and multiply them together to get your score. If the score fits one of the categories on the scorecard, write that score next to the category it matches. If it does not match a category, you score zero. If it fits more than one category, you can choose which category to use the score for. However, you can't switch that score to a different category later!

After one of the players has at least one score for all four categories, the players add up their scores. The player with the highest score wins.

You will need:
- 4 dice
- paper
- pencil.

Hint
Think about each category of number, and what possible combinations of dice rolls could produce that category. What strategy could you use to try to get as high a number as possible? How will you decide which dice to roll again on your second roll? Discuss your ideas as you play.

What did you learn?

Look back at the work you did in this chapter. Rate your progress.
1 = I cannot do this. **2** = I need more practice. **3** = I understand it and feel confident.

Can you:
- write numbers to one million in digits and number names and in expanded notation?
- use place value and value to compare numbers up to one million using the <, > and = signs?
- identify multiples and factors of numbers?
- work out whether a number is prime or composite?
- round numbers to the nearest thousand?

Review: Numbers, place value and rounding

Key terms and concepts

1 a You can represent the number 143 801 like this:

143 801 = (1 × 100 000) + (4 × 10 000) + (3 × 1000) + (8 × 100) + (0 × 10) + (1 × 1)

This is called ____ notation.

b You can also write it like this:

one hundred and forty-three thousand, eight hundred and one

This is called the ____ name.

2 a If you multiply 4 by 5, the ____ is 20.   **b** 4 and 5 are both ____ of 20.

c ____ numbers are all multiples of 2.   **d** ____ numbers end in 1, 3, 5, 7 or 9.

3 A ____ number can only be divided by itself and 1 without leaving a remainder. Examples are 5, 7 and 13.

Quick check

1 Write the number name for each number:

a 10 000   **b** 100 000   **c** 1 000 000

2 Write the number name for the total if you make ten groups of:

a 10   **b** 1000   **c** 100 000

3 Write the number 682 356 in expanded notation without brackets.

4 Round each number to the nearest thousand.

a 472 540   **b** 509 489

5 Write the following numbers in ascending order.

387 200   889 428   871 152   900 288   445 999   9999

6 The prime factors of a number are 2, 3, 5 and 7. What is the smallest number it could be?

7 List:

a all the factors of 24   **b** the first ten multiples of 12

Challenge and investigate

1 This table shows the population of six Caribbean countries in 2024. Use it to answer the questions that follow.

Saint Kitts and Nevis	47 755
Belize	410 825
Barbados	281 996
Saint Lucia	180 251
Dominica	73 040
Bahamas	412 624

a Which of these six countries has the greatest population?

b Which of these six countries has the smallest population?

c Round the smallest population to the nearest thousand.

d Write the population of Barbados using expanded notation.

2 Do your own research on the internet. Identify a Caribbean island that falls into each category:

a Population smaller than 25 000

b Population between 60 000 and 100 000

c Population closest to one million

3 a Draw a factor tree for the number 340.

b Express it as a product of all its prime factors.

SECTION 1

Chapter 2 Number patterns and number relationships

In this chapter, you will:
- use algebraic thinking to recognise, describe and explore number patterns
- use patterns to understand different kinds of numbers and solve problems
- solve algebraic problems
- solve number sentences involving one unknown.

Starting point

Discuss the picture.

1. What do you notice? What do you wonder?
2. What number patterns do you notice in the rows and columns?
3. How can you work out the numbers of the boxes that are not shown?

Key maths idea

A **number pattern** is a list of numbers in a particular order, that follows a **rule**.
Look at this number pattern: 0, 200, 400, 600, 800 …
The first **element** in this pattern is 0.
To get from each term to the next term, you use a rule.
In this pattern, the rule is add 200. We can use the rule to work out the next term.
800 + 200 = 1000

Key words
number pattern
rule
element

1. For each number pattern, identify the rule. Then say the element that comes next.

   a

   b

   c

78

d 17 27 37 47 ?

e 24 20 16 12 ?

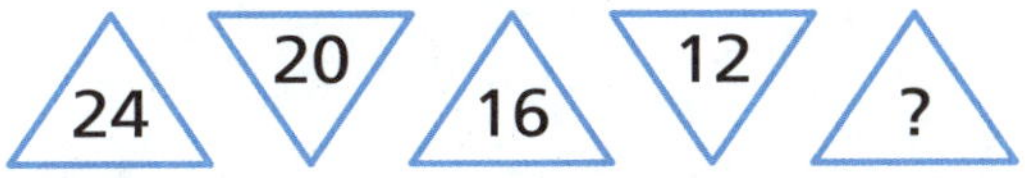

f 10 100 10 100 10 ?

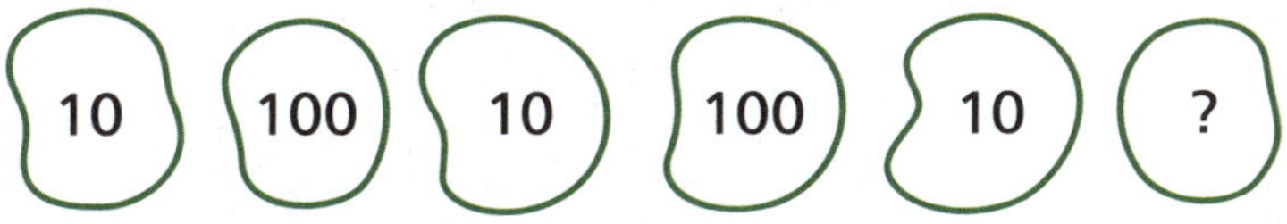

2   Make up a number pattern of your own. Ask a partner to work out the next element in the number pattern.

Repeating patterns

Key maths idea

To 'repeat' means to happen or appear again and again.
A **repeating pattern** has numbers or shapes that repeat in a regular way.

The **core** is the smallest unit that repeats to make the pattern. Once you find the core, you can work out the **element** that comes next. Look at this shape pattern:

Key words
repeating pattern
core
element

△○○△○○△○

In this pattern, the core is:

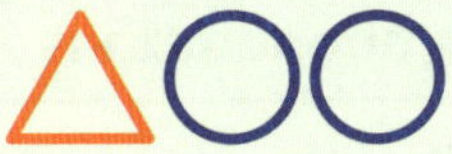

△○○

The next three elements must be:

○△○

Here are two repeating number patterns:
- 50, 100, 50, 100, 50 …   The core is: 50, 100

The next three elements must be 100, 50, 100
- 90, 100, 110, 100, 90, 100 …   The core is: 90, 100, 110, 100

The next three elements must be 110, 100, 90

1   Find the core of each pattern. Then draw the three shapes that come next.

a □|||□|||□   b T⊥T⊥T⊥

c  ↑→↓←↑→   d △▽□△▽□△

2   Find the core of each number pattern. Then write the three numbers that come next.
   a   5, 50, 5, 50, 5, …
   b   1, 10, 100, 1, 10, 100, 1, …
   c   2, 7, 3, 2, 7, 3, 2, …
   d   20, 40, 6, 20, 40, 6, 20, …

3   Identify the error in each of the patterns. Say which number should replace the error in each pattern.
   a   100, 200, 100, 200, 300, 200
   b   40, 80, 40, 100, 40, 80
   c   99, 77, 99, 66, 99, 77
   d   12, 24, 36, 12, 24, 12

Section 1 **Number** Chapter 2 Number patterns and number relationships

Real-life maths

Patterns are all around you. Look at these pictures. Can you see the patterns in these photos? Are there any other real-life patterns you can think of?

Increasing and decreasing number patterns

Key maths idea

When you look at a pattern, think:
- What is the first element?
- How does each element change from the one before?

Type of pattern	Description	Examples	Type of operations
Increasing	The numbers or values get bigger with each element	15, 30, 45, 60 … 1, 2, 4, 8, 16 …	Addition or multiplication
Decreasing	The numbers or values get smaller with each element	78, 77, 76 … 1000, 500, 250 …	Subtraction or division

For example: 7, 14, 21, 28, 35 …

The first element is 7. The rule is: add 7

You can see that the numbers get greater with each element. This tells you that it is an increasing pattern.

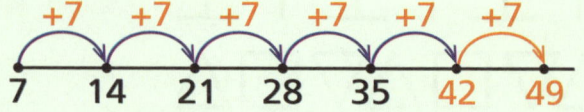

Key words
increasing
decreasing

Hint
Once you have guessed the rule, check whether it works in the same way for each element.

1   For each pattern, write I for increasing, D for decreasing or R for repeating. Then write the next three elements in the number pattern.

    a 10, 20, 30, 40, 50, 60, …
    b 7, 7, 5, 7, 7, 5, 7, 7, 5, …
    c 350, 300, 250, 200, …
    d 40, 80, 120, 160, …

Explore patterns in different kinds of numbers

2 a Press these keys on your calculator, and write the resulting number pattern.

[5] [+] [5] [=] [=] [=]

b Keep pressing the [=] key until your pattern has ten elements. Write them down.

Hint
This works on the calculator apps on a smartphone.

c Copy and complete the rule for the pattern you have made:
Start at ____ and ____ ____ to get the next element.

d Is it an increasing, decreasing or repeating pattern? How do you know?

3 Use your calculator to make a decreasing pattern with ten number elements. Write it down. Ask a partner to work out the rule you used.

4 Write the missing elements from each list.
  a 12 17 ____ 27 32 ____
  b 250 ____ 230 220 210 ____
  c 890 894 ____ 902 ____ 910
  d 81 72 ____ ____ 45 36

5 Each sentence has one element that is incorrect. Write which one is incorrect, and what the correct element should be in its place.
  a 40 80 100 160 200
  b 750 700 650 500 550

6 Write a number pattern with one element incorrect, for a partner to solve.

Explore patterns in different kinds of numbers

Key maths idea

When you divide an **even** number by 2, it leaves no remainder. When you count in twos, each number is even.

Even numbers: 2, 4, 6, 8, 10, 12, 14, 16, 18, 20 …

An **odd** number is always 1 more or 1 less than an even number. When you divide an odd number by 2, it leaves a remainder of 1.

Odd numbers: 1, 3, 5, 7, 9, 11, 13, 15, 17, 19 …

When you multiply a factor by itself, you get a **square number**. The factor you multiplied is the **square root**.

1 × 1 = 1   1 is the square root of 1
2 × 2 = 4   2 is the square root of 4
3 × 3 = 9   3 is the square root of 9

Key words
even   odd
square number
square root

1 Draw the next two shapes in the pattern of square numbers.

2 Copy and complete the table on the right about the pattern of square numbers.

3 What makes the pattern of square numbers different from the increasing patterns you have already worked with?

4 Continue the number pattern of square numbers up to twelve elements.

Element	How many pegs?
1	1
2	4
3	9
4	
5	

Section 1 Number Chapter 2 Number patterns and number relationships

5 a Look at the pattern of even numbers. Use it to complete the rule:
The ones digit of an even number is always one of these numbers:
____, ____, ____, ____ or ____.

b Look at the pattern of odd numbers. Use it to complete the rule:
The ones digit of an odd number is always one of these numbers:
____, ____, ____, ____ or ____.

6 Predict whether the answer will be odd or even for each of the following operations. Write the reason for your prediction. Then use a calculator to test out three or more examples. Were you right? Why?

a Add an even number to an even number.

b Subtract an odd number from an odd number.

c Subtract an even number from an odd number.

d Add an odd number to an odd number.

e Add three odd numbers together.

Multiplication patterns

> ### Key maths idea
>
> This is a multiplication chart. It shows the times tables up to 12 × 12.
>
> You can find many number patterns in the chart. For example, multiples in the 5 times table have a repeating pattern in the ones place.
>
> 5 times table: 5, 10, 15, 20, 25, 30, 35 …
>
> Pattern in the ones place: 5, 0, 5, 0, 5, 0, 5 …
>
> | × | 1 | 2 | 3 | 4 | 5 | 6 | 7 | 8 | 9 | 10 | 11 | 12 |
> |----|----|----|----|----|----|----|----|----|-----|-----|-----|-----|
> | 1 | 1 | 2 | 3 | 4 | 5 | 6 | 7 | 8 | 9 | 10 | 11 | 12 |
> | 2 | 2 | 4 | 6 | 8 | 10 | 12 | 14 | 16 | 18 | 20 | 22 | 24 |
> | 3 | 3 | 6 | 9 | 12 | 15 | 18 | 21 | 24 | 27 | 30 | 33 | 36 |
> | 4 | 4 | 8 | 12 | 16 | 20 | 24 | 28 | 32 | 36 | 40 | 44 | 48 |
> | 5 | 5 | 10 | 15 | 20 | 25 | 30 | 35 | 40 | 45 | 50 | 55 | 60 |
> | 6 | 6 | 12 | 18 | 24 | 30 | 36 | 42 | 48 | 54 | 60 | 66 | 72 |
> | 7 | 7 | 14 | 21 | 28 | 35 | 42 | 49 | 56 | 63 | 70 | 77 | 84 |
> | 8 | 8 | 16 | 24 | 32 | 40 | 48 | 56 | 64 | 72 | 80 | 88 | 96 |
> | 9 | 9 | 18 | 27 | 36 | 45 | 54 | 63 | 72 | 81 | 90 | 99 | 108 |
> | 10 | 10 | 20 | 30 | 40 | 50 | 60 | 70 | 80 | 90 | 100 | 110 | 120 |
> | 11 | 11 | 22 | 33 | 44 | 55 | 66 | 77 | 88 | 99 | 110 | 121 | 132 |
> | 12 | 12 | 24 | 36 | 48 | 60 | 72 | 84 | 96 | 108 | 120 | 132 | 144 |

Multiplication patterns

Mental maths

1. Look at the multiplication chart again. What pattern do you notice:
   a. as you go along the rows from left to right?
   b. as you go down the columns from top to bottom?

Hint
Look for digits that repeat, especially in the tens and ones places.

2. Is it possible that any of the numbers in the grid are prime numbers? Tell your partner your reason for your ideas.

3. What patterns do you notice in:
   a. multiples of 3, 6 and 9?
   b. multiples of 5 and 10?
   c. multiples of 2, 4 and 8?
   d. multiples of 11?

4. If you choose any number in the chart, how would you check whether it is a square number?

5. Choose a number in the 1 times table. Double it. What do you notice about the position of the double:
   a. if you look for it in the same row, but in a different column?
   b. if you look for it in the same column, but in a different row?
   Try it with another number in the 1 times table. What is the same, and what is different? Why?

Hint
To 'double' means to multiply by 2. To find half means to divide by 2.

6. Now repeat question **5** with the 2 times table. What do you notice?

---

1. Is each statement true or false? Use the multiplication chart to help you explain your answer.
   a. All multiples of 10 are also multiples of 5.
   b. A multiple of an even number must always be an even number.
   c. A multiple of an odd number must always be an odd number.
   d. The product of an odd and an even number must be an odd number.

2. Work out the missing elements in these number patterns.
   a. ____, ____, 30, 36, ____, ____, 54, 60
   b. ____, ____, 110, 99, ____, 77, ____, ____
   c. 21, 24, 27, ____, ____, ____
   d. 90, ____, ____, 63, ____, 45, 36

3. In each of the following number patterns, one of the elements is incorrect. Identify the mistakes and write the correct list of numbers in the pattern.
   a. 10, 15, 15, 25, 30, 35, 40
   b. 56, 63, 72, 80, 88, 96
   c. 24, 48, 70, 84, 108, 132

4. These patterns use doubling or halving. Write D for doubling or H for halving, and work out the next three elements in each number pattern.
   a. 2, 4, 8, 16, …
   b. 50, 100, 200, …
   c. 160, 80, 40, …
   d. 12, 24, 48, …

Section 1 Number Chapter 2 Number patterns and number relationships

> **Key maths idea**
>
> When you need to add, subtract, multiply or divide, you can look for **compatible numbers**. Some people also call these 'friendly numbers'.
> They are pairs of numbers that are easy to work with, for example:
> - pairs of numbers that add up to 10, 20 or 100
> - multiples of 5 or 10.
>
> **Key words**
> compatible numbers

1 a Write five different number sentences with a sum of 10.

  b Use your answer to question **a** to help you write five different number sentences with a sum of 100, and five more with a sum of 1000.

2 Use your answers from question **1**. Work out the missing numbers.

  a 35 + ____ = 100
  b 100 − ____ = 23
  c ____ + 79 = 100
  d 100 = ____ − 90

Using patterns to solve problems

> **Key maths idea**
>
> Tara is having a party. She wants to give each child 2 lollipops.
> There will be 12 children at the party.
> How many lollipops must she get?
> Look at how these two students solved the problem.
>
>
>
> **Adara**
>
> | Number of children | 1 | 2 | 3 | 4 | 5 |
> |---|---|---|---|---|---|
> | Number of lollipops | 2 | 4 | 6 | 8 | 10 |
>
> The first element is 2. The rule is add 2. To find the 12th element, I keep adding 2. 2, 4, 6, 8, 10, 12, 14, 16, 18, 20, 22, 24 Tara must get 24 lollipops.
>
> **Jenny**
>
> 1 child   2 lollipops
> 2 children   4 lollipops
> 3 children   6 lollipops
>
> I can see that the number of lollipops is double the number of children.
> Double 12 = 24
> Tara must get 24 lollipops.

1. Tara wants to make sure that there are 3 patties for each child at the party.
   a. What number pattern can she use to work out how many patties to buy?
   b. Patties come in boxes of 10. How many boxes should Tara buy? Show your working.
2. Larry says that if a pattern rule is 'add zero', this is the same as a pattern rule of 'multiply by 1'.
   a. Test Larry's rule. Does it work? Why?
   b. What happens if the pattern rule is 'subtract zero'?
3. Yanick says that if the pattern rule is 'multiply by zero', then you always get the same repeating pattern, no matter the starting element. Test Yanick's idea. Does it work? Why?
4. Deborah is building a pattern using toothpicks. She uses three toothpicks for the first shape, five toothpicks for the second shape and seven toothpicks for the third shape.
   a. What rule does her pattern follow?
   b. How many toothpicks would she need for the eighth shape in the pattern? Show how you work it out.
   c. Draw a set of shapes that would follow this pattern rule.
   d. Compare your work with a partner. What similarities and differences do you notice?

Finding an unknown

Key maths idea

Example 1

Look at this number sentence:

3 +  = 10

Think: 'three plus something equals ten'
You know that 7 + 3 = 10.

So ★ = 7

Key words

unknown
bar model
number line
decomposition
number facts

The star is like a question mark or an answer box. It represents an amount or value we do not know and we want to find out. We call it an **unknown**.

Example 2

23 + ☐ = 45

Look at how these different students worked out the unknown.

Ashanti used a **bar model** to show the problem as a picture.

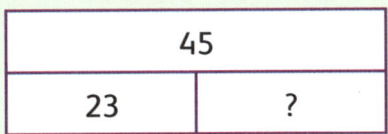

Randy thought about jumps on a **number line**. He started at 23. He used smaller jumps to get to 45. This is called **decomposition** method, because it decomposes (breaks down) a number into smaller parts.

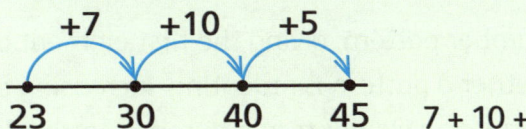

7 + 10 + 5 = 22

This helps you to understand the **number facts** you can use to solve it:

23 + ☐ = 45
45 − ☐ = 23
45 − 23 = ☐

```
   45
 − 23
 ─────
   22    So 23 + 22 = 45
```

25

Section 1 Number Chapter 2 Number patterns and number relationships

1 Work out the unknown in each number sentence. Explain how you worked it out, or show your working.

a ☐ + 20 = 50   b 2 × ☐ = 100   c ☐ − 25 = 340   d double ☐ = 1000
e ☐ ÷ 3 = 6   f 700 − ☐ = 200   g 50 + ☐ = 90   h 20 ÷ ☐ = 4

2 Dinesh tells Erica he will sell twice as many raffle tickets as she does.

a If Erica sells 18 tickets, how many must Dinesh sell?

b If Erica sells 35 tickets, how many must Dinesh sell?

c Explain the procedure you can use to work out how many tickets Dinesh must sell if you know how many Erica sold.

3 For each of the following sentences, the result is 100. Write a number sentence for each. Use a box, shape or letter to represent the unknown number. Then work out the unknown.

a I choose a number and double it.

b I choose a number and add 35.

c I choose a number and divide it by 10.

d I choose a number and subtract 80.

Problem solving

Use what you know about different types of numbers, number facts or compatible numbers. Solve these problems.

1 Use the digits 3, 4, 5 and 6. Copy and complete the addition sum using all four digits each time to make:

a the smallest possible sum

b the greatest possible sum.

2 Work out the missing digit for each box.

a
```
  4 1 ☐
−  ☐ 1 0
─────────
  1 0 2
```

b
```
   ☐ 2 0
−  3 1 ☐
─────────
   2 0 1
```

c
```
   ☐ 4 5
−  6 ☐ 3
─────────
     6 2
```

Hint
Think about what you would need to trade from the next column.

What did you learn?

Look back at the work you did in this chapter. Rate your progress.
1 = I cannot do this. **2** = I need more practice. **3** = I understand it and feel confident.

Can you:
- describe a number pattern, giving the first element and the pattern rule?
- work out whether a pattern is repeating, increasing or decreasing, and explain why?
- work out the missing elements in a number pattern?
- create repeating, increasing and decreasing number patterns of your own?
- describe patterns in different kinds of numbers, including even, odd and square numbers?
- describe patterns that appear in a list of multiples?
- use patterns to solve problems?
- find unknowns in number sentences?

Review: Number patterns and number relationships

Key terms and concepts

1 Write the word that correctly completes each statement.

| element | increasing | number pattern | core | repeating | rule | decreasing |

  a A list of numbers in a particular order is known as a ____.
  b In the pattern 4, 8, 12, 16, 20, … the ____ is add four.
  c In the pattern above, 4 is the first ____.
  d This is an example of a ____ pattern: 5, 6, 5, 6, 5, 6. The ____ is 5, 6.
  e The first element of a pattern is 100, and the rule is subtract 3. It is a ____ pattern.
  f Adding or multiplying usually results in an ____ pattern.

Quick check

1 Give an example of:
  a an increasing pattern   b a decreasing pattern   c a repeating pattern.
2 Choose one of the multiplication tables. Write the first six multiples in a table. Then complete the statement of the pattern rule:
   Start at ____ and ____ ____ to get to the next element.
3 Write the first ten square numbers.
4 Work out the value of the unknown in each equation.
  a ☐ × 10 = 800   b 700 − ☐ = 20   c 180 + ☐ = 180

Challenge and investigate

1 In a pattern, the first element is 25, the fifth element is 225 and the fourth element is 175. Write the first ten elements in the pattern.
2 The first four elements of a pattern are shown here:

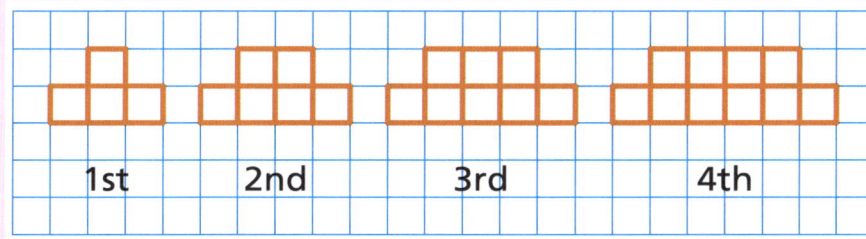

1st   2nd   3rd   4th

  a State the pattern rule.
  b How many squares will make up the tenth element of this pattern?
3 Lara is building model stairs out of toy cubes. She follows this pattern:
   How many cubes will Lara need to build the next model, with five steps?
   Show how you work it out.

1 step   2 steps   3 steps   4 steps

SECTION 1

Chapter 3 Number operations

In this chapter, you will:
- create and solve problems using the four operations
- develop and apply mental mathematics strategies
- use estimation to check and justify your answers
- use inverse operations
- find approximate solutions to problems that do not require an exact answer.

Starting point

1. The masses of these four sweet potatoes are:
   369 g, 378 g, 399 g and 382 g.

   a Is it fairer to round their masses to 300 g or to 400 g? Why?

   b Which number facts could you use to estimate the mass of the four together?

   c How many sweet potatoes would you expect to get in a 5 kg bag?

The four operations

Key maths idea

Addition, subtraction, multiplication and division are **operations**. Each operation has its own symbols.

Key word: operation

Operation and symbol	Words we use	Useful to remember
Addition **+**	sum total how many altogether	You can add in any order. 2 + 3 = 3 + 2
Subtraction **−**	difference how many / much more how many fewer how much less	Subtraction is the inverse of addition. 3 + 2 = 5 5 − 3 = 2 5 − 2 = 3
Multiplication **×**	product times	You can multiply in any order. 2 × 3 = 3 × 2 Multiplying by 1: 2 × 1 = 2 Multiplying by 0: 2 × 0 = 0
Division **÷**	divide find the quotient how many times does it go into …	Division is the inverse of multiplication. 3 × 4 = 12 12 ÷ 3 = 4 12 ÷ 4 = 3 Dividing by 1: 5 ÷ 1 = 5 You cannot divide by 0

1. Decide which operation to use, then work out the answer.
   a the sum of 200, 1000 and 50
   b the product of 100 and 10
   c how many times 20 goes into 120
   d the total if you add 19, 190 and 3000
   e the difference between 2000 and 800
   f five times fifty
   g 200 more than 1900
   h 800 less than 2000

Talking maths

Think about this question:
How many times can you fill a 1 litre bucket from a 12 litre tank?
It uses the word 'times', but you would solve it with another operation. Explain why that is.

Compatible numbers

Key maths idea

When you work with operations, look for **compatible numbers** (sometimes called friendly numbers). They are numbers that are easy to work with, such as:
- pairs of numbers that make 10 (such as 6 and 4, or 7 and 3)
- numbers that end in one or more 0s (such as 10, 200, 4000)
- numbers that end in 5
- facts that you already know, such as doubles or halves.

Key words
compatible numbers

Mental maths

1. Say or write:
   a five addition sentences that make 100, using multiples of 10
   b five addition sentences that make 1000, using multiples of 100

2. Say or write:
   a double 5   double 50   double 500   double 5000
   b double 2   double 20   double 200   double 2000
   c half of 8   half of 80   half of 800   half of 8000
   d half of 18   half of 180   half of 1800

3. Add:
   a 7 + 3   b 17 + 13   c 170 + 130   d 1700 + 1300

4. Explain which pair of compatible numbers you could use for all the sums in question **3**.

5. Subtract:
   a 13 – 5   b 130 – 50   c 1300 – 500   d 50 – 20
   e 50 – 25   f 500 – 250   g 5000 – 2500

6. Explain which pair of compatible numbers you could use for all the sums in question **5**.

Section 1 Number Chapter 3 Number operations

1. Add up each sum. Use compatible numbers to help you add.
   - **a** 90 + 70 + 20 + 80 + 10 + 30
   - **b** 120 + 190 + 50 + 80 + 150 + 10
   - **c** 130 + 95 + 270 + 197 + 5 + 3
   - **d** 540 + 80 + 60 + 50 + 220 + 50

> **Key maths idea**
>
> You can use number facts you know to work with bigger numbers easily. For example, it is easy to multiply big numbers by 10. You can use this to help you multiply big numbers by 9 or 11.
>
> **Example 1**
> 175 × 9 = ?
> 175 × 10 = 1750
> Now I need to take away one set of 175
> 1750 − 175
>
>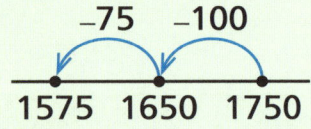
>
> **Example 2**
> 345 × 11 = ?
> 345 × 10 = 3450
> Now I just add another set of 345
> 3450 + 345 = 3795
> 345 × 11 = 3795

1. Use compatible numbers or number facts to help you work out the answers.
   - **a** 145 × 11
   - **b** 194 × 9
   - **c** 435 × 11
   - **d** 207 × 9

2. Use the inverse operation to check your answers to question 1.

Mental methods

> **Key maths idea**
>
> A mental calculation is one that you work out in your head, rather than using pen and paper to write out every step of your working in full. Sometimes mental strategies include jotting notes or steps while you work things out in your head. Here are some strategies to help you calculate mentally.
>
> **Use number facts or patterns you already know:**
>
> 48 × 100 = ?
>   Think: 48 × 1 = 48
>   So 48 × 100 = 4800
>
> 5 × 400 = ?
>   Think: 5 × 4 = 20 so 5 × 400 = 2000
>
> **Use split strategy:**
>
> 73 + 24 = ?
>   Think: 70 + 20 = 90
>   73 + 24   97
>   3 + 4 = 7
>
> **Use doubling and halving:**
>
> 412 × 2 = ?
>   Think: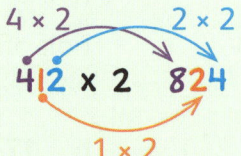
>
> 44 × 5 = ?
>   Think:
>
>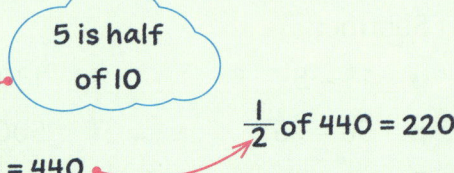
>
> 44 × 10 = 440
> 5 is half of 10
> $\frac{1}{2}$ of 440 = 220

Estimating with rounding and compensation

Problem solving

Use any of the written or mental strategies you have learned to solve these problems. Use estimation to help you judge whether your answers are reasonable.

1. A travel company offers holiday packages to Trinidad and Tobago. Customers can book at a cost of $1445 per person per day, or $10 000 per week.
   a. How much does a three-day holiday cost if a customer books on the daily rate?
   b. What savings does the weekly rate offer compared to the daily rate?

2. A hotel charges $1268 per room per night. What would the total be for a visitor who stays for 5 nights?

3. A theatre charges $85 for tickets. The table shows how many tickets are sold for all the performances in a week.

Tuesday	Wednesday	Thursday	Friday	Saturday
189	124	215	360	360

   a. On which days were there no performances?
   b. Which were the most popular days for tickets?
   c. Did the theatre earn more from the weeknight shows or the weekend shows? How can you work it out without doing the full calculations?
   d. Estimate the total earnings for the week.

Extension

4. The theatre in question **3** has a **minimum** number of tickets that they must sell each night, otherwise the show is cancelled. On Wednesday, the number was three times the minimum. What is the minimum number of tickets?

Key word
minimum

Estimating with rounding and compensation

Key maths idea

Before you solve a problem, you can **estimate** the answer. Estimating means guessing about how much the answer will be. You work out an **approximate** answer.

Key words
estimate
approximate
front-end rounding
compensate

Example 1

364 + 280 + 555 + 662
↓
300 + 200 + 500 + 600

This is called **front-end rounding**. You round each number using only the number at the front, and zeros for all the other places.

300 + 200 + 500 + 600 ≈ 1600   The ≈ sign means approximately equal to.

Rounding in this way does not give a very accurate estimate.

We can **compensate** by rounding some of the numbers up and some down.

With compensation:

300 + 200 + 600 + 700 ≈ 1900   Instead of rounding them all down, we round two numbers down and two numbers up.

Accurate sum with a calculator: 364 + 280 + 555 + 662 = 1861

31

Section 1 **Number** Chapter 3 Number operations

> **(continued)**
>
> **Example 2**
> 1984 − 762
> With front-end rounding: 1000 − 700 ≈ 300
> With compensation: 2000 − 700 ≈ 1300
> Accurate sum with a calculator: 1984 − 762 = 1222
>
> **Key word**
> minuend
>
> When you subtract, you should usually round the **minuend** (the number from which you subtract) up, not down. Can you work out why?

1. Round each number, then estimate the sum.
   a 49 + 28 + 22 + 69 + 55
   b 899 + 264 + 467 + 551
   c 1825 + 2950 + 1509 + 2671

2. Check your answers to question 1.

3. Use front-end rounding to estimate the difference between the numbers below. Then use a calculator to work out the accurate answer.
   a 749 − 376   b 2948 − 1153   c 8031 − 2348

4. Check your answers to question 3 using the inverse operation.

Remember: subtraction and addition are inverse operations.

Addition and subtraction problems

> **Key maths idea**
>
> When you solve problems using addition or subtraction:
> - estimate to help you work out what a reasonable answer would be
> - if you do not need an exact answer, just use the estimate to solve the problem
> - use inverse operations to check your answer.
>
> **Example 1**
>
> Jemila is a farmer. She has 1524 banana trees, 984 guava trees and 658 mango trees.
> How many trees are there altogether?
> Estimate: 2000 + 900 + 600 = 3500
> Written working out:
>
> ```
>    Th  H  T  O
>        2  1  1
>     1  5  2  4
>        9  8  4
>  +     6  5  8
>  =  3  1  6  6
> ```
>
>
>
> - Make sure to line up the place values.
> - Work from right to left.
> - When a column sums to more than 9, regroup.
> - Compare the answer to the estimate. Does it look reasonable?

Addition and subtraction problems

(continued)

Example 2

Jemila wants to plant more guava trees so their numbers are equal to the banana trees.
How **many more** guava trees must she plant?

Estimate: 1500 − 900 = 600

```
    Th H T O
     0 14 1
     1̸ 5̸ 2 4
  −    9 8 4
  =    5 4 0
```

- The number you are subtracting from (the minuend) goes on top.
- Make sure the place values line up.
- Work from right to left.
- If the bottom digit is too great to subtract, regroup.
- Check the calculation against your estimate. Does it look reasonable?

1 Calculate the sum of each set of numbers. Show any regrouping.

a   Th H T O
     4 1 5 9
     2 3 7 6
   + 1 1 1 8

b   Th H T O
     2 4 8 5
     3 1 1 2
   + 1 2 3 9

c   Th H T O
     1 6 5 8
     1 3 7 5
   + 1 0 4 4

d   Th H T O
     2 3 0 9
     1 0 2 8
   + 2 3 8 7

e   Th H T O
     1 0 4 8
     2 1 3 6
   + 4 4 5 5

f   Th H T O
     1 0 2 3
     3 4 9 7
   + 1 0 0 9

g   Th H T O
     2 0 7 3
     1 6 4 9
   + 2 0 1 8

h   Th H T O
     3 0 4 0
     2 2 3 7
   + 1 0 8 6

2 Subtract. Set out your work as shown. Show any regrouping.

a   Th H T O
     7 8 4 5
   − 2 3 8 7

b   Th H T O
     9 0 1 2
   − 1 0 5 5

c   Th H T O
     8 3 0 7
   − 4 0 4 9

d   Th H T O
     4 5 1 4
   − 2 3 8 8

e   Th H T O
     9 7 0 8
   − 5 4 9 4

f   Th H T O
     7 0 8 5
   − 4 9 3 9

g   Th H T O
     7 3 9 2
   − 5 0 1 6

h   Th H T O
     6 4 8 3
   − 4 0 7 6

3 Estimate a reasonable answer. Then calculate.

  a What is the sum of 3852 and 4397?
  b What is the difference between 3852 and 4397?
  c A driver travelled 1438 km in one week and 1079 km in the second week. How far did he travel altogether?
  d How much further did the driver travel in the first week than in the second?

4 Check your answers to question **3** using inverse operations.

Section 1 **Number** Chapter 3 Number operations

Problem solving

1. Shania rubbed out part of her homework by mistake. Work out the missing numbers.

   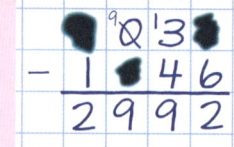

   a) 2■64 + 145 = 250■

   b) ■6■8 + 1063 = 8721

   c) ■⁹0¹3■ − 1■46 = 2992

   d) 3■¹³4■ − 1055 = 2188

2. Make up your own example of an addition or subtraction sum with smudged numbers. Exchange with a partner and solve each other's problems.

   Hint
   Use inverse operations and number facts you already know to help you.

3. The table shows the distances by plane between Port of Spain and the capitals of some other island countries in the Caribbean. You do not need to work out exact answers.

   a. Which capital is furthest from Port of Spain?
   b. Which is closest to Port of Spain?
   c. A pilot flew to Basse-Terre and back. Approximately how far did she fly?
   d. In a week, a plane flies to Caracas and back, to Kingstown and back, and to Castries and back. Approximately how much is the total distance?
   e. A pilot flies to one of these capitals and back. She flies a distance of almost 1000 km. Which capital city was it?
   f. Make up a question of your own about the table to ask a partner.

Capital city	Distance from Port of Spain by plane
Basse-Terre, Guadeloupe	782 km
Bridgetown, Barbados	341 km
Caracas, Venezuela	618 km
Castries, St Lucia	407 km
Fort-de-France, Martinique	491 km
Kingstown, St Vincent and the Grenadines	305 km
Oranjestad, Aruba	994 km
Roseau, Dominica	513 km
Willemstad, Curaçao	850 km

   Hint
   Remember: if a flight will go there and back, it will go twice the distance!

Multiplication strategies

Key maths idea

When you multiply, you can estimate using multiplication facts you already know.

3521 × 3 = ?

Think: 3000 × 3 = 9000
       4000 × 3 = 12 000

So 3521 × 3 should be between 9000 and 12 000

Multiplying by a 2-digit number

(continued)

Partitioning

3521 × 3
= (3000 × 3) + (500 × 3) + (20 × 3) + (1 × 3)
= 9000 + 1500 + 60 + 3
= 10 563

- Expand the number you want to multiply.
- Find the product of each bracketed multiplication.
- Find the total.

Use place value

```
      ¹3 5 2 1
  ×         3
  ―――――――――
    1 0 5 6 3
```

- Put the greater number at the top.
- Work from right to left (from the smallest to greatest place value).
- Regroup if the answer is a 2-digit number.

1 Estimate the answer. Then use any method to multiply. Check the reasonableness of your answer using your estimate.

a 933 × 5   b 398 × 3   c 496 × 5   d 2413 × 3
e 1582 × 4   f 3075 × 2   g 4047 × 2   h 2165 × 4

Problem solving

1 In a parking garage, 365 cars can fit on a full floor. How many cars are parked if the garage has:
   a 2 full floors?   b 5 full floors?
2 Parking costs $7 per hour. What is the total received if 2 floors are full for 3 hours?

Multiplying by a 2-digit number

Key maths idea

- Long multiplication is a way of multiplying bigger numbers.
- Start with an estimate.
- Write the greater number on top, and the smaller number underneath.
- Place digits under the correct place value positions.
- Multiply each digit in the bottom number with each digit in the top number.
- Work from smallest place value (ones), to greatest place value.
- Add the products at the end.

Example 1

52 × 19 = ?
Estimate:
50 × 20 = 1000

```
      H T O
        ¹5 2
  ×      1 9
  ―――――――――
        4 6 8
  +     5 2 0
  ―――――――――
        9 8 8
```

52 × 19 means 19 groups of 52
This is the same as 10 groups of 52 plus another 9 groups of 52
In long multiplication, start with the ones 9 × 52 = 468
Multiply the tens.
10 × 52 = 520
Then add them together: 468 + 520 = 988
52 × 19 = 988

Section 1 **Number** Chapter 3 Number operations

> **(continued)**
>
> **Example 2**
> 765 × 23 = ?
> Estimate: 800 × 20 = 16 000
>
> ```
>   HTh Th  H  T  O
>          ¹7 ¹6  5
>    ×         2  3
>       2   2  9  5
>   + 1 5   3  0  0
>     1 7   5  9  5
> ```
>
> **Example 3**
> 4013 × 12 = ?
> Estimate: 4000 × 10 = 40 000
>
> ```
>   HTh Th  H  T  O
>        4  0  1  3
>    ×         1  2
>        8  0  2  6
>   + 4  0  1  3  0
>     4  8  1  5  6
> ```

1 Estimate, then use long multiplication to find the product of each number sentence.

   a 31 × 14   **b** 36 × 18   **c** 47 × 14   **d** 87 × 34
   e 271 × 32   **f** 816 × 43   **g** 187 × 49   **h** 378 × 96
   i 1273 × 11   **j** 2038 × 25   **k** 2920 × 51   **l** 4709 × 74

Exploring multiplication strategies

> **Key maths idea**
>
> Mariah, Mia and Kimani each used a different method to solve the same question.
> A driver earns $26 per hour. He works 49 hours in a week. How much does he earn altogether?
> 49 × 26 = ?
>
> **Mariah**
>
> Expanded numbers
>
> ```
>      40 + 9
>   × 20 + 6
>
>   6 × 9   =   54
>   6 × 40  =  240
>   20 × 9  =  180
>   20 × 40 =  800
>              1274
> ```
>
> **Mia**
>
> In a grid
>
>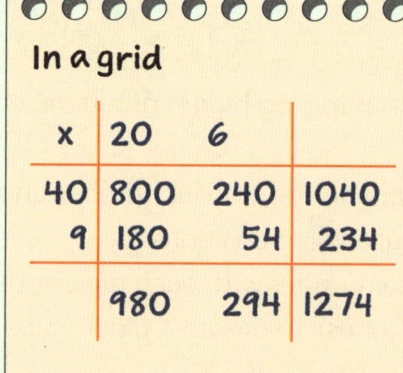
>
> **Kimani**
>
> Long multiplication
>
>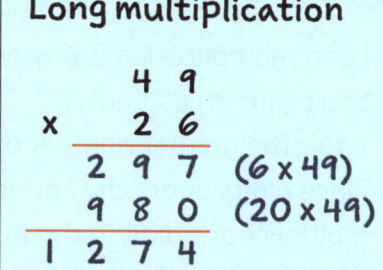

1 Discuss the examples above with a partner.
   • What do you notice that is similar about their working?
   • How can you tell they all got the answer right?
   • Which rules about multiplication did they use?
   • Which method would you choose to use? Why?

2 Write the missing number that makes each statement correct.
   a 145 × 32 = (145 × 2) + (145 × ☐)
   b 7842 × 41 = (7842 × 40) + (☐ × 1)
   c 3055 × ☐ = (3055 × 20) + (3055 × 4)
   d ☐ × 83 = (1607 × 3) + (1607 × 80)

3 Look at Reza's working to solve this multiplication question.

   27 × 15 = ?

   27 × 15
   ╱╲ ╱╲
   ③×⑨×③×⑤

   = 3 × 3 × 5 × 9

   = 9 × 45

   10 × 45 = 450

   so 9 × 45 = 450 − 45

   = 405

   Discuss with a partner:
   • How did Reza make the numbers easier to work with?
   • Does it always work to rearrange the factors in a multiplication sentence? Try a few examples of your own to test it out.
   • How did Reza use the 10 times table to work out a multiplication by 9?
   • Can you think of a different way to solve this problem?

The relationship between multiplication and division

Key maths idea

Division involves **equal sharing**. We use the ÷ sign to show division.
Division is the **inverse** of multiplication. You can use multiplication facts to work out related division facts.

Key words
equal sharing
inverse

Example

When you multiply a whole number by 10, 100 or any other multiple of 10, you get 0s in the product. This can help you with division and multiplication by bigger numbers.

6 × 7 = 42	60 × 7 = 420	60 × 70 = 4200	600 × 70 = 42 000
42 ÷ 7 = 6	420 ÷ 60 = 7	4200 ÷ 60 = 70	42 000 ÷ 70 = 600

1 Make notes of the multiplication facts you can use to help you.
   a 2000 ÷ 10   b 2000 ÷ 20   c 2000 ÷ 50   d 7000 ÷ 70
   e 2000 ÷ 40   f 1200 ÷ 30   g 1500 ÷ 30   h 1500 ÷ 50
   i 350 ÷ 70   j 4500 ÷ 90   k 2700 ÷ 90   l 4800 ÷ 60

Section 1 **Number** Chapter 3 Number operations

Full STEAM ahead Working with a calculator

You can use a calculator to perform calculations and to check your working.

This diagram shows all the features on a simple calculator. Compare it to the calculators in your classroom.

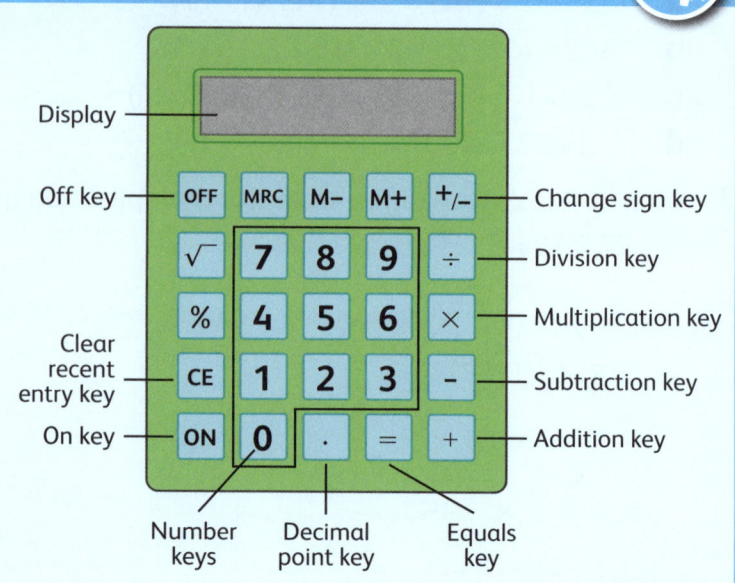

Work in groups.

- One student writes a 4-digit number on a piece of paper, without the others looking. This will be the target number.
- Another student enters a 2-digit number into the calculator before they look at the target number.
- The group looks at both numbers.
- Discuss how you can multiply the 2-digit number by a single-digit number to get as close as possible to the 4-digit number. You can use repeated multiplication by different whole numbers. The aim is to reach the number (or as close as possible) in as few steps as possible. Keep notes of the steps.
- If another student thinks they can do it in fewer steps, let them try with a different number.
- You can change the game so that you explore division. In this version, the first student writes a 4-digit starting number, and the second student notes a single-digit target number. Everyone tries to reach that number with as few single-digit divisions as possible.

Long division

Key maths idea

When you talk about division, it is useful to remember the different names for the numbers in the division sentence.

1500 ÷ 50 = 300   1500 is the **dividend**.
50 is the **divisor**.   300 is the **quotient**.

When the divisor is a multiple of 10, it is easy to work out the answer mentally. Some divisions are not easy to work out mentally.

Estimate using rounded numbers. Then use **long division**. This is a written method that helps keep track of each step in your working.

Key words
dividend
divisor
quotient
long division

Example

There will be 335 guests at a wedding. The organiser has ordered 22 tables. How many people will fit at each table?

335 ÷ 22 = ?   Estimate: 340 ÷ 20 = ?   34 ÷ 2 = 17 so 340 ÷ 20 = 17
        This estimate tells us around 17 people will fit at a table.

Multi-step problems

(continued)

Step 1:

```
        1
   _____
22 | 3 3 5
    -2 2 ↓
    ─────
      1 1 5
```

Start with the biggest place value (hundreds): 3 hundreds
22 does not go into 3. Move to the next place (tens).
22 goes into 33 once (think: 22 × 1 = 22; 22 × 2 = 44)
Write the 1 above the tens.
Multiply 1 × 22. Write the product underneath.
Subtract to get the remainder. 33 – 22 = 11. We have a remainder of 11 tens.
The red arrow shows that we subtract the next place (ones).

Step 2:

```
        1 5 r 5
   _____
22 | 3 3 5
    -2 2 ↓
    ─────
      1 1 5
     -1 1 0
     ─────
           5
```

Work in a similar way.
How many times can 22 go into 115?
22 × 4 = 88. That is too small.
22 × 6 = 132. That is too high.
22 × 5 = 110
Write the 5 over the ones.
Multiply 5 × 22 = 110. Write the product underneath.
Subtract to get the remainder. 115 – 110 = 5

335 ÷ 22 = 15 remainder 5

15 people will fit at each table, but there are 5 more people, so five of the tables will have 16 people, and the rest will have 15.

1 Divide the following sums. Use the long division method you learned here.

- **a** 98 ÷ 15
- **b** 74 ÷ 12
- **c** 83 ÷ 11
- **d** 93 ÷ 13
- **e** 148 ÷ 13
- **f** 287 ÷ 15
- **g** 4089 ÷ 12
- **h** 1576 ÷ 18
- **i** 9047 ÷ 14
- **j** 3817 ÷ 27
- **k** 5496 ÷ 41
- **l** 8090 ÷ 45

2 12 people shared a meal at a restaurant. The total came to $2820. How much should each person pay if they share the bill equally?

3 Janice sold raffle tickets for $23 each. She collected a total of $1587. How many tickets did she sell?

4 A farmer picked 8476 oranges. She packaged them into bags of 15 oranges. How many bags did she fill, and how many oranges were left over?

Multi-step problems

Key maths idea

Sometimes you need to do more than one operation to solve a problem.

Example

Maggie buys old furniture and sells it to make a **profit**.

Key word
profit

Profit is the difference between what something cost and what you sell it for. You will explore this in more detail in Chapters 12 and 15.

Section 1 **Number** Chapter 3 Number operations

(continued)

Maggie paid $4200 for two tables. She sold one for $3480 and the other for $3640. What was her profit for the two tables?

Think: Her profit is what she sold them for, minus what she paid.

Sold for:

$\quad$ $ 3 4 8 0
+ $ 3 6 4 0
$\quad$ $ 7 1 2 0

Minus what she paid:

$\quad$ $ 7 1 2 0
− $ 4 2 0 0
$\quad$ $ 2 9 2 0

Her profit was $2920.

Problem solving

For each problem, read it first for understanding. Decide which operations will help you to solve it. Also think about how many calculations you will need to do.

1. Maggie bought an armchair for $3290 and a small table for $890. She sold them as a set for $7000. What was her profit on the set?

2. Maggie also makes handmade candles. She makes $12 profit per candle, and she earns about $900 per month in total from the candles. About how many candles does she sell each year?

3. Four students earned these marks on their exams. Each mark is out of a total of 100.

Name	English	Mathematics	Science
Billy	76	82	72
Keshon	75	81	79
Onika	81	79	77
Josiah	74	78	76

The total for all three subjects is a mark out of 300. Who got the highest total?

What did you learn?

Look back at the work you did in this chapter. Rate your progress.
1 = I cannot do this. **2** = I need more practice. **3** = I understand it and feel confident.

Can you:
- solve problems involving addition, subtraction, multiplication and addition?
- calculate mentally using compatible numbers, number facts, doubling and halving, or any other strategies?
- estimate to check or justify an answer?
- use inverse operations to check your answers or to calculate?
- find approximate solutions for questions that do not require an accurate answer?

Review: Number operations

Key terms and concepts

1. Complete each sentence with the correct operation: addition, subtraction, multiplication or division.
   a. Addition is the inverse operation of ____.
   b. Division and ____ are inverse operations.
   c. You use ____ to work out the product of two numbers.
   d. You use ____ to work out the difference between two numbers.
   e. ____ can mean equal sharing.

2. Fill in the blanks.
   ____ rounding means rounding to the greatest place value in a number, for example rounding 1840 equals 1000. This is different from rounding to the nearest thousand, which would round 1840 to ____.

Quick check

1. Calculate:
   a. the sum of 1700, 450 and 48
   b. the difference between 1800 and 750
   c. the product of 1000 and 50
   d. double 1850

2. Calculate mentally:
   a. 1500 + 2300
   b. 7000 × 15
   c. 1800 ÷ 30
   d. 1500 ÷ 5
   e. 1500 ÷ 50
   f. 6400 ÷ 80

3. Use long division or long multiplication to calculate:
   a. 138 × 23
   b. 1248 ÷ 11
   c. 2819 × 35
   d. 9847 ÷ 42

Challenge and investigate

1. Nigel bought 23 pencils for 45 cents each. What did he pay?
2. Vanessa runs 15 kilometres every week.
   a. How far will she run in a year?
   b. How far will she run in ten years?
3. Each classroom in a school has 28 chairs. If the school has 124 classrooms, how many chairs are there?
4. How many banana plants are there in 25 rows if there are 45 plants in each row?
5. Nesha did a multiplication on the whiteboard. Two digits got rubbed out by mistake. What are they?

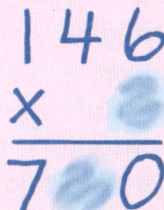

6. Mrs Huggins saved $12 948 in a year. She saved an equal amount each month of the year. What was the monthly amount she saved?

SECTION 2

Chapter 4 Angles

In this chapter, you will:
- identify different kinds of lines and angles
- identify horizontal, vertical, parallel and perpendicular lines
- explore the relationship between right angles and different kinds of turns: quarter-turns, half-turns, three-quarter turns and full turns
- solve problems involving turns and angles.

Starting point

1. Discuss these pictures with a partner.
   a. Where do the lines cross each other?
   b. Which lines form triangles?
   c. What other shapes do the lines form?

Different kinds of lines

Key maths idea

A **point** is a position in space. When we join two points, we create a **line**. Shapes are made out of lines. We use some special words to describe lines.

——————— straight

This is a **straight** line.

Squares and triangles have straight sides.

| vertical
|_____ horizontal

Vertical lines go up and down, like a flagpole or a lamppost. **Horizontal** lines run from left to right, like the horizon.
Perpendicular lines meet at a right angle. The corner of a square or a rectangle is formed by perpendicular lines.

 curved

Some lines are **curved**.

Circles and ovals have curved lines.

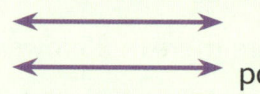 parallel

A pair of lines is **parallel** if they run next to each other without meeting. Think of the sides of a ladder, or train tracks. The opposite sides of squares and rectangles are parallel.

Key words
point
line
straight
curved
vertical
horizontal
perpendicular
parallel

42

Angles

1 Draw:
   a   a pair of horizontal parallel lines
   b   a pair of vertical parallel lines
   c   two different pairs of perpendicular lines.

2 Draw a picture of a car or a boat. Label examples of straight, curved, parallel and perpendicular lines on your picture.

Full STEAM ahead   Model your maths terms

1 Glue seeds on card to illustrate each term:
   a   a point
   b   a straight line
   c   a curved line
   d   perpendicular lines
   e   parallel lines

You will need:
- seeds or beans
- glue
- paper or card
- ruler
- pencil.

Angles

Key maths idea

When two lines meet or cross at a point, the meeting point forms the **vertex** of an **angle**. An angle is a measure of **turn**.

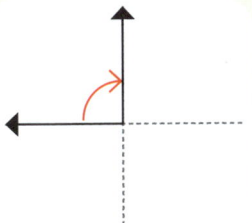

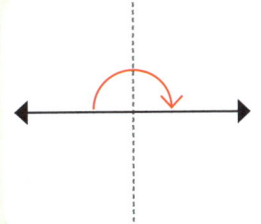

A **quarter-turn** forms a right angle. This is equal to the angle formed where two perpendicular lines meet.

A **half-turn** is the same as a straight line. It is equal to two quarter-turns.

Key words
vertex
angle
turn
quarter-turn
half-turn
three-quarter turn
full turn

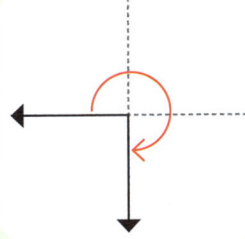

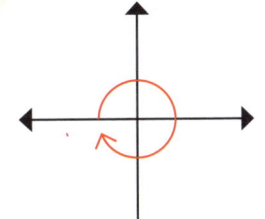

A **three-quarter turn** is made of three right angles.

A **full turn** is also called a revolution. It is equal to two half-turns or four right angles.

Section 2 Geometry Chapter 4 Angles

1 The arrow marks an angle. In each set, identify the angle that is not the same size as the others. Explain how you decided.

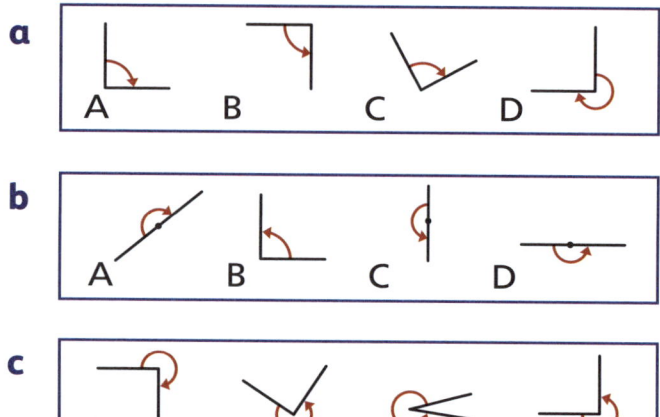

2 Here is a set of mixed angles. Copy the table. Write the letters of the angles in the correct rows.

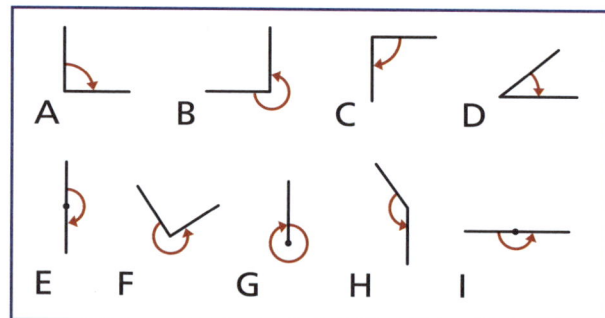

Type of angle	Letter that matches
quarter-turn	
half-turn	
three-quarter turn	
full turn	
none of the above	

Full STEAM ahead   Make models of angles

1 Fasten strips of card, or wooden or plastic strips, to make geo-strips.

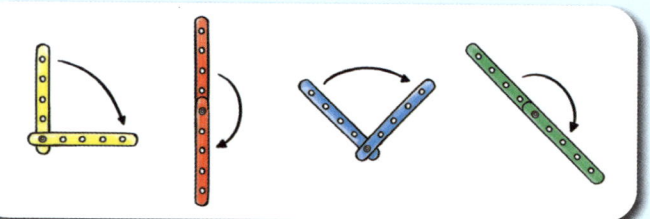

You will need:
- strips of card
- split pins or paper fasteners
- pairs of scissors.

2 With a partner, look around at objects in your classroom that form angles.

3 Use your geo-strips to make angles of different sizes that you find around you.

4 Discuss:
 a Does the length of the arms change the size of the angle? Why or why not?
 b What changes the size of the angle?

5 Find objects in your classroom that have these angles:
 a a right angle
 b a straight line
 c a three-quarter turn
 d a full turn

44

Turning clockwise and anticlockwise

> **Key maths idea**
>
> The time on this clock is quarter past seven. At 7 o'clock, the minute hand pointed to the 12. It took 15 minutes to make a quarter-turn in a **clockwise** direction.
>
> On a working clock, the hands always turn clockwise. The long hand will make another three-quarter turn by 8 o'clock.
>
> The opposite to clockwise is **anticlockwise** or **counterclockwise**.
>
> We use these terms to tell us the direction and size of a turn. A turn can also be called a **rotation**.
>
>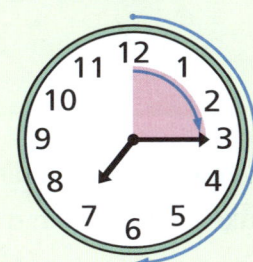
>
> **Key words**
> clockwise
> anticlockwise
> counterclockwise
> rotation

1 For each angle, write the size of turn (full, half, quarter or three-quarter) and the direction (clockwise or anticlockwise).

a  b 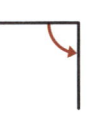 c  d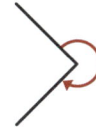

e  f 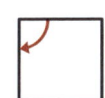 g  h

2 What size of turn does the minute hand of a clock make in:

a 30 minutes?
b 15 minutes?
c 45 minutes?

3 For each clock, write the size of the turn (quarter, half, three-quarter or full) the minute hand makes as the time passes:

a from 11:00 to 11:15
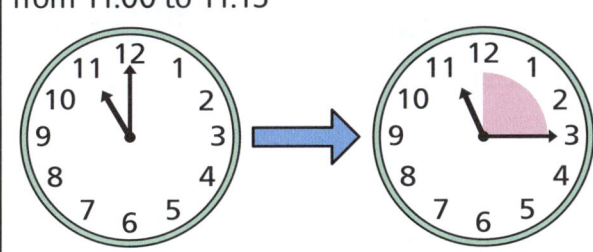

b from 3:15 to 3:45
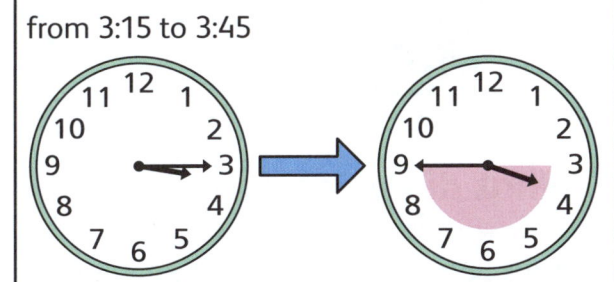

c from 1:20 to 1:35
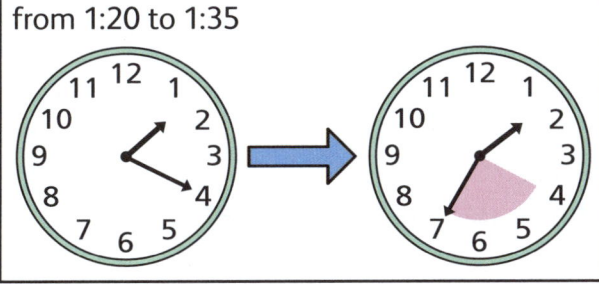

d from 11:10 to 11:55
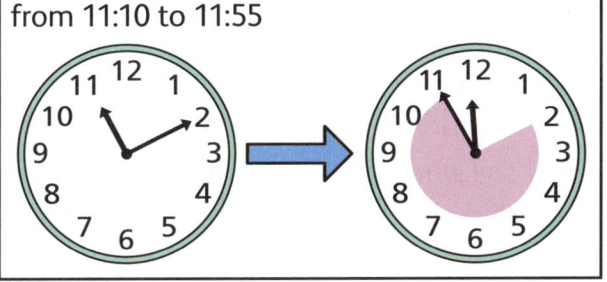

Section 2 Geometry Chapter 4 Angles

Turns in different directions

> **Key maths idea**
>
> We use cardinal directions to help us find our way.
> The main cardinal directions are north (N), west (W),
> east (E) and south (S).

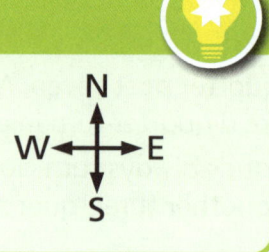

Draw a sketch to help you.

1. For each set of directions, work out which direction you will end up facing.
    a. Start facing north. Make a half-turn in a clockwise direction.
    b. Start facing west. Make a quarter-turn in a clockwise direction.
    c. Start facing east. Make a three-quarter turn anticlockwise.

2. Work out the starting positions if you:
    a. make a three-quarter turn clockwise, and end up facing east
    b. make a half-turn and end up facing north.

3. Write your own set of directions for a partner to follow.

> **Talking maths**
>
> 1. We use the word 'turn' in many different ways. Read each sentence and say what 'turn' means. Is it similar to or different from the mathematical meaning?

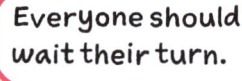

In folklore, Anansi can turn into many other creatures.

Everyone should wait their turn.

It is always possible to turn over a new leaf.

That is an opportunity you cannot turn down.

What did you learn?

Look back at the work you did in this chapter. Rate your progress.
1 = I cannot do this. **2** = I need more practice. **3** = I understand it and feel confident.

Can you:
- identify horizontal and vertical lines?
- say whether a pair of lines is perpendicular or parallel?
- describe the size of a turn as a quarter-turn, half-turn, three-quarter turn or full turn?
- solve problems involving clockwise and anticlockwise turns in different directions?

Review: Angles

Key terms and concepts

1. Write the term for each description.
    a A position in space, usually shown on a diagram as a dot
    b A pair of lines that are the same distance apart and never meet
    c A word we use to describe lines that are upright (go up and down)
    d The size of a quarter-turn
    e The straight path between two points

2. How many right angles make up:
    a a quarter-turn?   b a half-turn?   c a three-quarter turn?   d a full turn?

Quick check

1. You have learned four different ways to describe a square corner. Write them down.

2. Draw an example of:
    a a quarter-turn   b a half-turn   c a full turn.

3. What angle does the minute hand of the clock turn in:
    a half an hour?   b fifteen minutes?   c one hour?

4. Work out the position that each arrow is facing by the end of its movement. Draw your answer.
    a An arrow is pointing **up**. It turns one quarter-turn in an anticlockwise direction, then a three-quarter turn in a clockwise direction.
    b An arrow is pointing **left**. It makes a half-turn in a clockwise direction, then a quarter-turn anticlockwise.

Challenge and investigate

1. Imagine you are giving a visitor directions to get from your classroom to the school office. Use some of the terms you learned in this chapter to give the directions.

2. Here are some pictures of everyday things.

   a Write an example of parallel lines shown in each item in the pictures.
   b Write an example of perpendicular lines shown in each item in the pictures.
   c Look at the picture of the shelves. Explain why right angles are important in the shelf's design. In other words, how do the right angles help it to work?
   d Look at the picture of the slide. Why must the rungs of the ladder be parallel?
   e Draw pictures of an object at home or at school that uses parallel or perpendicular lines in its design. Make notes on your drawing about how the angles help the object to work properly.

> **Hint**
> Think about how the shelf would work if the brackets were a different angle.

SECTION 2

Chapter 5 Solids and plane shapes

In this chapter, you will:
- develop an understanding of the properties of solids and plane shapes
- solve problems involving solids and plane shapes
- explore the properties of triangles.

Starting point

1. Discuss the picture with a partner.
   a. Which of the **solid shapes** have you seen before?
   b. Name the **plane shapes** that make up their **faces**.

2. The shapes below are all **polygons**.

   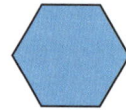  

   a. Do you remember what a polygon is? Use the shapes to help you write a definition.
   b. Check your definition with a partner or in a dictionary.
   c. What do you notice about the three polygons on the left and the three on the right?

3. Look around your classroom and find objects that match any of the solids or shapes in the pictures.

Key words
solid shape
plane shape
face
polygon

Angles in polygons

Key maths idea

A polygon is a flat shape with straight lines that form its **sides**. The sides meet at points called **vertices** (corners). The word 'polygon' comes from the Greek words 'poly' (many) and 'gon' (angle). Remember: when two lines meet at a point, they form an angle. This means the number of vertices is the same as the number of angles inside the shape. **Regular** polygons have all sides equal in length. They also have angles that are all the same size.

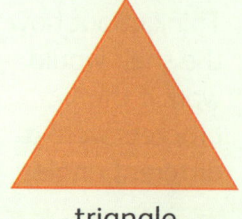
triangle

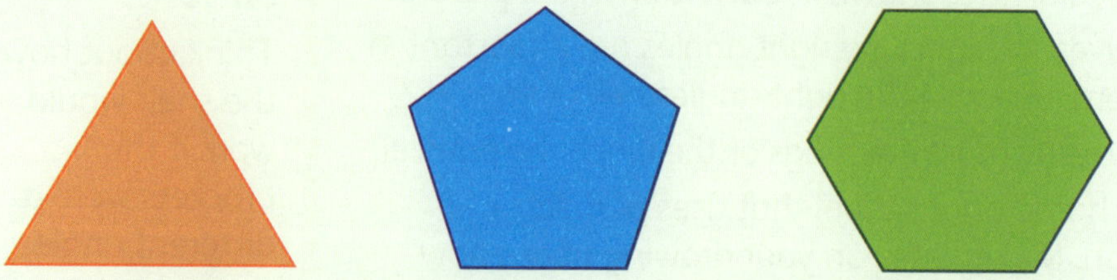

pentagon

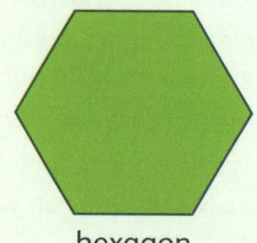
hexagon

Key words
sides
vertices (singular: vertex)
regular

48

Sorting and classifying triangles

1. Take a sheet of paper. The corners are right angles. Use a right angle to help you measure the angles in the regular triangle, pentagon and hexagon.

2. Use your paper angle. Investigate the angles inside these regular polygons.

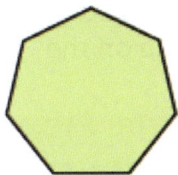 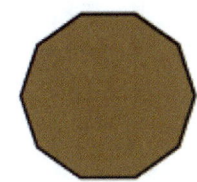 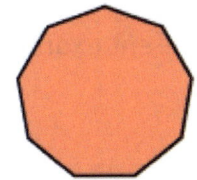 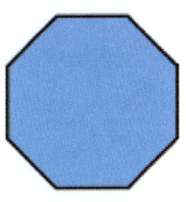

Key word
irregular

3. What pattern do you notice about the number of sides in a regular polygon, and the size of its inside angles?

4. Here are some **irregular** shapes.
   a. Use your square angle to compare the size of the angles.
   b. How can you tell that the pattern you noticed in question 3 is not true for irregular shapes?

5. Use a ruler or set square to draw three pairs of parallel lines. Find ways to join the parallel lines to make:
   a. a square   b. a rectangle   c. a triangle.

   Explain what you had to do differently to make each shape.

6. Draw three more pairs of parallel lines. Draw different kinds of regular and irregular polygons that are not squares, rectangles or triangles.

Sorting and classifying triangles

Key maths idea

All triangles have three sides and three angles, but they can look very different. When we **sort** shapes, we put them into groups based on **properties** that they share. We have some special names for triangles, which depend on the answers to these questions:

- How many sides are the same length?
- How many angles are the same size?
- Does it have a right angle?

Key words
sort
properties
classify
right-angled triangle
equilateral
isosceles
scalene

Once you know the answers to these questions, you can **classify** triangles using the special names below.

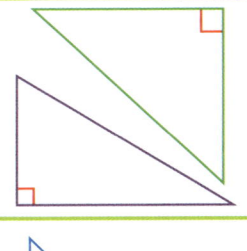

A **right-angled triangle** contains a 90° angle.

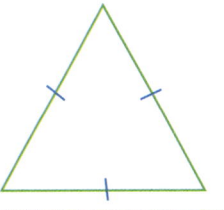

An **equilateral** triangle has all its sides equal in length, and all its angles equal in size.

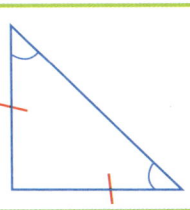

An **isosceles** triangle has two sides equal in length. The two base angles are equal in size.

Scalene triangles have all three sides different lengths, and all three angles different sizes.

49

Section 2 Geometry Chapter 5 Solids and plane shapes

1   Complete these questions either in pairs or individually.
   a   Draw lines to divide a page into four blocks.
   b   Label the blocks 'right-angled', 'equilateral', 'isosceles' and 'scalene'.
   c   In each block, draw at least three different examples of that type of triangle.

Hint
To draw equilateral triangles, use a folded sheet of paper to help you copy the angle from the example on the previous page.

Mental maths

1   Is it possible to make these triangles? Give reasons for your answers.
   a   an equilateral triangle that contains a right angle
   b   three scalene triangles put together to form a rectangle
   c   an isosceles triangle that contains a right angle
   d   a square built out of right-angled triangles

Hint
Draw sketches to show your thinking.

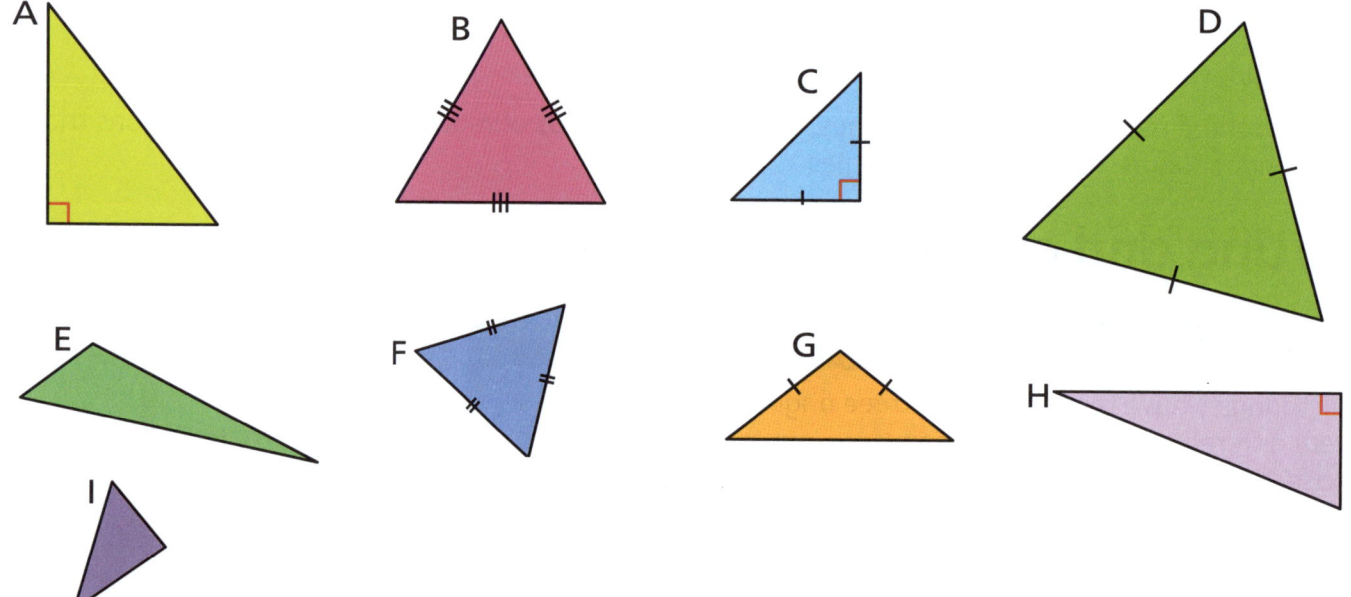

2   Use the set of triangles above. Write the letters of:
   a   the equilateral triangles
   b   the scalene triangles
   c   the isosceles triangles
   d   the right-angled triangles.

3   Identify something that is the same, something that is similar and something that is different about the triangles in each pair. Give a reason.
   a   triangles A and H
   b   triangles A and C
   c   triangles C and G

Faces of solids

> **Full STEAM ahead** Make a model
>
> Draw or make models of the angles and shapes you have learned about in this chapter. You may choose:
> - maths equipment such as geoboards
> - digital drawing using a computer or tablet
> - an online maths app
> - models using craft materials (for example, string, tape, pipe cleaners or any other materials)
> - upcycled recyclable items, such as plastic or cardboard.

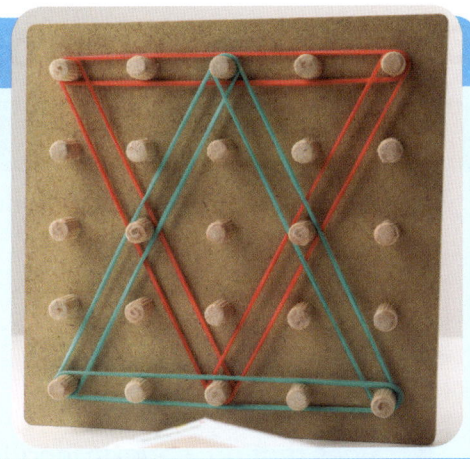

Faces of solids

> **Key maths idea**
>
> A cube is a solid. It is a **three-dimensional (3-D)** object. It has length, width and height. The flat surfaces of a solid are the faces. The line where two faces meet is called an **edge**. Edges meet at a point called a vertex. (The plural of vertex is vertices.) 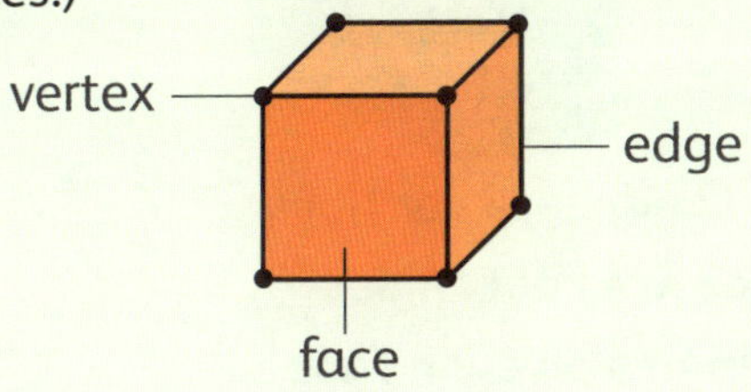
>
> Some solids have **curved** surfaces. They are not faces, because they are not flat shapes. 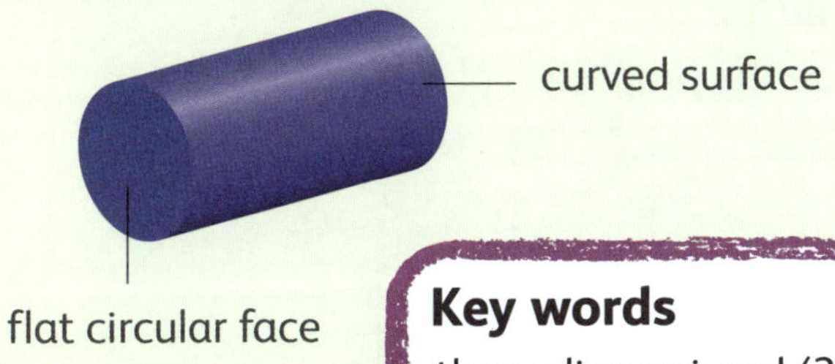
>
> curved surface
>
> flat circular face
>
> **Key words**
> three-dimensional (3-D)
> edge
> curved

1 Name the solid that has:
   a six square faces
   b no flat faces
   c two parallel triangular faces
   d four triangular faces
   e one round face
   f two parallel round faces.

2 Which four solids have parallel faces?

3 Which solids have edges that are perpendicular?

 cube
 cuboid
 cylinder
 cone
 triangular-based prism
 square-based pyramid
 sphere
 triangular-based pyramid

> **Real-life maths**
>
> Look at each of the solids above. Think of an example of each one from the real world around you.

Section 2 Geometry Chapter 5 Solids and plane shapes

Describing the properties of shapes and solids

Key maths idea

To **describe** the properties of plane shapes or solids, think about these questions:

Plane shapes	Solids
How many … • sides? • vertices? • sides that are equal in length? • equal angles? • right angles? • parallel sides? Remember: irregular shapes have sides of different lengths.	How many … • faces? • curved surfaces? • edges? • vertices? • parallel or perpendicular edges? What are the shapes of the faces? Are the faces regular or irregular shapes?

Key word
describe

Example 1
Describe the properties of this shape.

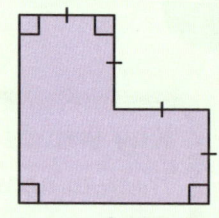

- irregular polygon (not all sides are equal lengths)
- 6 sides, 6 vertices
- 4 right angles
- 4 sides equal in length
- opposite sides are parallel

Example 2
Describe the properties of this solid.

- cuboid
- 6 faces, 12 edges, 8 vertices
- all 6 faces are rectangles
- 4 right angles on each face
- 2 pairs of parallel edges on each face
- perpendicular edges at all vertices
- opposite edges are parallel

1 Draw the faces of the following solids and describe their properties.
   a a cube
   b a square-based pyramid
   c a triangular prism

2 Identify the shape or solid that does not belong in each set. Give reasons.

a
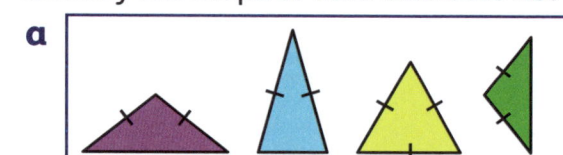

b
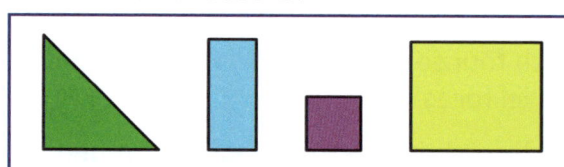

c

d

Describing the properties of shapes and solids

3 Name a solid that has:
   a eight vertices   b no vertices   c one vertex
   d five vertices   e four vertices   f six vertices.

4 How would you construct each of these shapes from cut-out cardboard? Imagine you need to cut out each face from cardboard, and the edges will be joined with tape or glue. Draw the shapes you would need to make all the faces of the following shapes. Use a ruler.
   a a cuboid
   b a cube
   c a triangle-based prism
   d a square-based pyramid
   e a cone
   f a cylinder

Hint
You can turn back to the 'Faces of solids' section to remind yourself of the names of the solids you have learned about.

5 Why is it not possible to make a sphere by cutting out faces from cardboard?

When you think about solids with curved surfaces, think about how they would look if you could unroll the surface to make a flat shape.

Talking maths

1 Discuss with a partner the similarities and differences between:
   a cubes and cuboids
   b triangular-based prisms and triangular-based pyramids
   c cones and cylinders.

What did you learn?

Look back at the work you did in this chapter. Rate your progress.
1 = I cannot do this. 2 = I need more practice. 3 = I understand it and feel confident.

Can you:
- name plane shapes and identify their properties?
- construct and draw regular and irregular polygons?
- identify parallel and perpendicular lines in different shapes?
- describe the angles in solids and plane shapes?
- classify triangles according to their sides and angles?
- solve problems involving solids and plane shapes?

Section 2 Geometry Chapter 5 Solids and plane shapes

Review: Solids and plane shapes

Key terms and concepts
Write the words missing from each statement.
1. We can classify triangles according to how many sides they have that are equal in length. The terms we use are:
   a ____: two sides equal in length
   b ____: three sides equal in length
   c ____: all sides different lengths
2. Write in your own words what each of the following terms mean. Draw a labelled diagram to explain.
   a face
   b edge
   c vertex

Quick check
1. Draw an example of:
   a a right-angled triangle
   b an isosceles triangle
   c an equilateral triangle
   d a scalene triangle.
2. Draw diagrams of the following solids. Make notes of the faces, edges and vertices.
   a a cube
   b a cuboid
   c a triangular prism
   d a square-based pyramid
   e a cylinder
   f a cone
3. Name each shape or solid described.
   a A four-sided plane shape with four right angles but some sides shorter than others
   b A plane shape with five sides
   c A plane shape with four sides of equal length and both pairs of opposite sides parallel
4. What are the four ways we classify triangles according to side length and angles? Draw diagrams.

Challenge and investigate
1. You have learned four different ways to describe a square corner. Write them down.
2. Which of these types of triangles cannot have a right angle: isosceles, scalene or equilateral? Draw a sketch to explain your answer.
3. What is the special name we give to:
   a a regular three-sided polygon?
   b a regular four-sided polygon?
   c a solid with six rectangular faces?
4. An isosceles triangle has one side 8 cm long and another side 4 cm long. What are the possible lengths of the third side?
5. Copy and complete this table about some of the solids you have learned about.

	Number of surfaces	Number of flat faces	Number of curved surfaces	Number of edges	Number of vertices
cuboid					
triangular prism					
square-based pyramid					
cone					
cylinder					

SECTION 3

Chapter 6 Fractions

In this chapter, you will:
- apply previous knowledge of fractions
- add, subtract, multiply and divide fractions
- solve problems involving fractions and the four operations.

Key words
whole
fraction
part
model
equivalent

Starting point

1. Look at the picture and answer these questions in pairs.
   a. How do you know that the round orange shape represents one **whole**?
   b. How many whole shapes (like the orange one) can you make by putting the **fractional** pieces in the picture together?

2. Mia says that you can make a whole by combining different sized **parts**. She draws this **model** to show her ideas: $\frac{1}{2} + \frac{1}{4} + \frac{1}{4} = 1\ whole$
   a. Mia says her model shows that one-half is **equivalent** to two-quarters. What does she mean?
   b. Draw your own models to show four different ways of making a whole using different-sized parts.

3. Vanessa says this is an increasing pattern: $\frac{1}{2}, \frac{1}{3}, \frac{1}{4}, \frac{1}{5}, \frac{1}{6}, \frac{1}{7} \ldots$
   a. Draw diagrams to show Vanessa why this is a decreasing pattern.
   b. A pattern starts with $\frac{2}{3}$ and each term is $\frac{1}{3}$ greater than the previous one. Write the first five terms in the pattern.

Section 3 **Number** Chapter 6 Fractions

Adding and subtracting fractions

Key maths idea

Do you remember how to add and subtract fractions with the same **denominator**?
These examples show you how you add or subtract fractions when the denominators are the same.

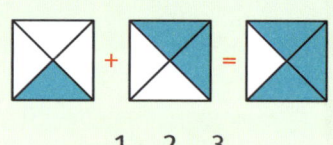

$\frac{1}{4} + \frac{2}{4} = \frac{3}{4}$

$\frac{1}{8} + \frac{4}{8} = \frac{5}{8}$

> **Hint**
> If you add eighths to eighths, your answer will also be in eighths.
> To find the total number of eighths, you can just add the **numerators**.

Sometimes your answer might be an **improper fraction**.

$\frac{6}{5} + \frac{3}{5} = \frac{9}{5}$

$\frac{3}{4} + \frac{4}{4} = \frac{7}{4}$

You can convert improper fractions to equivalent **mixed numbers**.

$\frac{9}{5}$ is the same as $\frac{5}{5} + \frac{4}{5} = 1\frac{4}{5}$

$\frac{7}{4}$ is the same as $\frac{4}{4} + \frac{3}{4} = 1\frac{3}{4}$

> **Hint**
> You must be able to convert between improper fractions and mixed numbers.
> $\frac{14}{5} = \frac{5}{5} + \frac{5}{5} + \frac{4}{5} = 2\frac{4}{5}$
> $3\frac{1}{4} = \frac{4}{4} + \frac{4}{4} + \frac{4}{4} + \frac{1}{4} = \frac{13}{4}$

$\frac{5}{6} - \frac{2}{6} = \frac{3}{6}$

$\frac{8}{10} - \frac{6}{10} = \frac{2}{10}$

$\frac{3}{6}$ and $\frac{2}{10}$ are not in their **simplest form**. A fraction is in its simplest form when no number greater than 1 can divide into both the numerator and the denominator.

To **simplify** a fraction, divide the numerator and the denominator by the same number. This is sometimes called reducing the fraction.

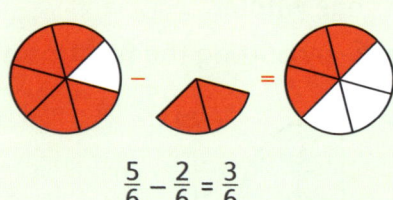

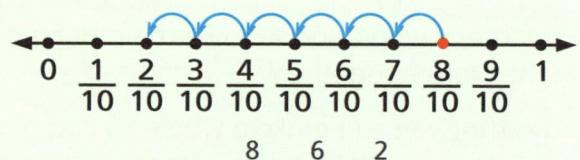

Key words
denominator
numerator
improper fraction
mixed number
simplest form
simplify

Whole numbers and fractions

1 Do these calculations mentally.
   a $\frac{2}{5} + \frac{1}{5}$
   b $\frac{3}{10} + \frac{4}{10}$
   c $\frac{3}{7} + \frac{2}{7}$
   d $\frac{3}{8} + \frac{2}{8}$
   e $\frac{5}{6} - \frac{1}{6}$
   f $\frac{7}{8} - \frac{3}{8}$
   g $\frac{3}{5} - \frac{3}{5}$
   h $\frac{9}{7} - \frac{5}{7}$

2 How many tenths are there in $1\frac{7}{10}$? How could you write that as an improper fraction?

3 Maggie says that $\frac{3}{10} + \frac{5}{10}$ is $\frac{8}{20}$. What has she done wrong?

4 Look at this pattern: $\frac{1}{2}, \frac{2}{4}, \frac{3}{6}, \frac{4}{8}, \frac{5}{10}, \frac{6}{12} \ldots$
   a Describe the pattern in your own words.
   b Write the next three terms in the pattern.

Problem solving

1 Alexia spent $\frac{2}{9}$ of her money on groceries, $\frac{1}{9}$ on rent and $\frac{4}{9}$ on transport.
   a What fraction of her money did she spend?
   b What fraction of her money does she have left?

2 Marcus has $\frac{6}{7}$ of a metre of rope and Josiah has $\frac{5}{7}$ of a metre of rope.
   a How much longer is Marcus's piece of rope?
   b They need to mark out a length of 2 m. How much more rope will they need? Show how you work this out.

3 What is the perimeter of a square if each side is $\frac{3}{10}$ of a metre long?

Whole numbers and fractions

Key maths idea

You can add fractions to whole numbers. Your answer can be a whole number or a mixed number.

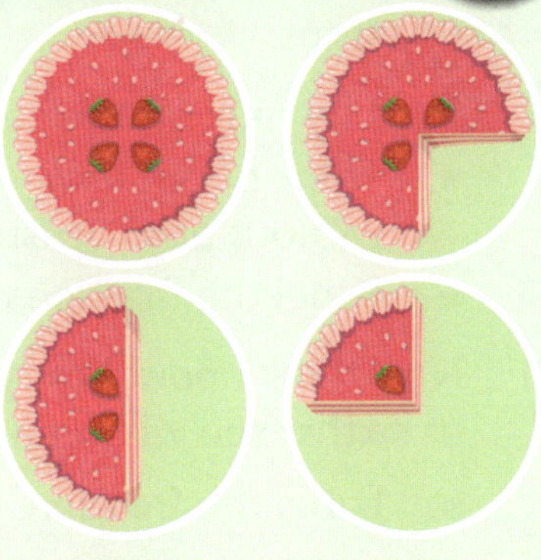

The picture shows a whole cake, $\frac{3}{4}$ of a cake, $\frac{1}{2}$ a cake and $\frac{1}{4}$ of a cake. These pieces can be added together to get different amounts of cake.

- 1 whole cake plus $\frac{3}{4}$ of a cake gives you $1\frac{3}{4}$ cakes
- 1 whole cake plus $\frac{1}{2}$ a cake gives you $1\frac{1}{2}$ cakes
- 1 cake + $\frac{1}{4}$ of a cake = $1\frac{1}{4}$ cakes
- 1 cake + $\frac{3}{4}$ of a cake + $\frac{1}{4}$ of a cake = 2 cakes

How much cake is there altogether in the picture?

You can also subtract fractions from whole numbers. It may help to write the whole number as a fraction. How much pie is left if you eat one slice?

The pie is divided into eighths. You can see that one whole pie is equivalent to $\frac{8}{8}$.

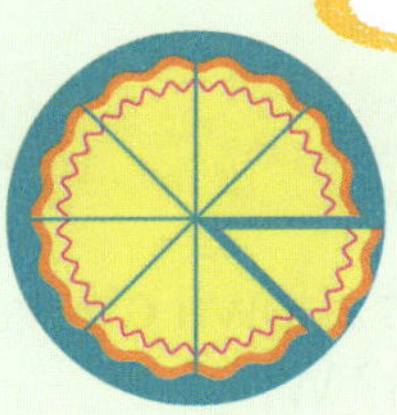

$\frac{8}{8} - \frac{1}{8} = \frac{7}{8}$ of a pie left

Hint
Any fraction with an equal numerator and denominator is equal to one whole.
$\frac{2}{2} = \frac{3}{3} = \frac{4}{4} = \frac{5}{5} = 1$

Section 3 Number Chapter 6 Fractions

1. Add these whole numbers and fractions. Draw diagrams or models if you need to.
   - a $3 + \frac{2}{3}$
   - b $12 + \frac{5}{6}$
   - c $9 + \frac{1}{2}$
   - d $\frac{1}{3} + 7$
   - e $4 + \frac{1}{4}$
   - f $9 + \frac{2}{3}$
   - g $15 + \frac{9}{11}$
   - h $4 + \frac{19}{20}$
   - i $3 + \frac{2}{3}$
   - j $4 + \frac{4}{9}$
   - k $2 + \frac{6}{6}$
   - l $4 + \frac{9}{10}$

2. Subtract these whole numbers and fractions. Draw diagrams or models if you need to.
   - a $1 - \frac{9}{10}$
   - b $1 - \frac{1}{2}$
   - c $1 - \frac{3}{4}$
   - d $1 - \frac{5}{9}$
   - e $3 - \frac{2}{3}$
   - f $2 - \frac{1}{2}$
   - g $10 - \frac{1}{10}$
   - h $5 - \frac{4}{9}$
   - i $2 - \frac{1}{4}$
   - j $3 - \frac{4}{9}$
   - k $10 - \frac{9}{10}$
   - l $4 - \frac{4}{4}$

3. How much pizza is on the table? Show how you work this out.

> **Full STEAM ahead** Make an equivalent fractions chart
>
> Use strips of paper to make a set of equivalent fraction strips.
>
> **You will need:**
> - A4 paper, coloured paper or Bristol board
> - ruler and markers
> - scissors or a craft knife.
>
> 1. Prepare the strips:
>   - Draw lines across the width of the page. Try to make these approximately the same height.
>   - Cut or tear along the lines to make a set of at least eight strips.
> 2. Start by making this set of equivalent fractions.
>   - Think about how you can fold the strip to make each set without any measuring.
>   - Label the fractions on each strip and set them aside.
> 3. Now think about how you can make a strip of thirds, and one of sixths. You will need to measure to do this.
>   - Label the fractions and add these strips to the others.
> 4. How can you make fifths and tenths? Share your ideas in pairs and then make a strip of fifths and one of tenths.
>   - Put all the strips in order of fraction size to make a chart like the one in the picture. This will help you find equivalent fractions in the next section.

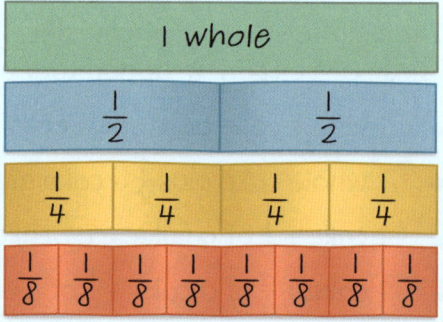

> **Talking maths**
>
> Look at the halves, quarters and eighths shown above. What happens to the size of the fraction as the denominator gets bigger? Why?

Working with different denominators

Key maths idea

When you add or subtract fractions with different denominators, you can use equivalent fractions to make the denominators the same. Use your fraction strips to help you.

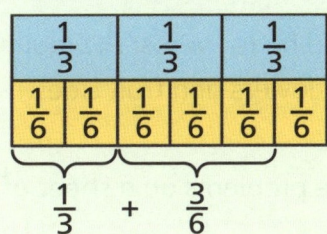

$\frac{1}{3} = \frac{2}{6}$

$\frac{2}{6} + \frac{3}{6} = \frac{5}{6}$

Add $\frac{1}{3} + \frac{3}{6}$

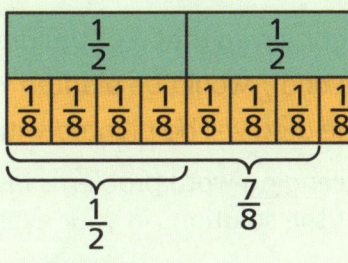

$\frac{1}{2} = \frac{4}{8}$

$\frac{7}{8} - \frac{4}{8} = \frac{3}{8}$

Subtract $\frac{7}{8} - \frac{1}{2}$

You can also multiply or divide to make equivalent fractions. Remember: you must multiply and divide both the numerator and the denominator by the same number.

Add $\frac{2}{9} + \frac{1}{3}$

$\frac{2}{9} + \frac{1}{3} \times \frac{3}{3}$ Convert thirds to ninths

$= \frac{2}{9} + \frac{3}{9}$

$= \frac{5}{9}$

Subtract $\frac{3}{4} - \frac{2}{12}$ Give your answer in its simplest form

$\frac{3}{4} \times \frac{3}{3} - \frac{5}{12}$ Convert quarters to twelfths

$= \frac{9}{12} - \frac{5}{12}$

$= \frac{4}{12}$ This is not in its simplest form, so reduce the fraction

$= \frac{4}{12} \div \frac{4}{4} = \frac{1}{3}$

Talking maths

1 a Explain why you need to have the same denominators to add or subtract fractions.
   b How do you decide which fraction to convert so the denominators are the same?
   c Share your ideas in groups.

1 Find the sum of each pair of fractions. Use fraction strips or draw your own models if you need to.

a $\frac{1}{4} + \frac{3}{8}$ b $\frac{1}{3} + \frac{1}{6}$ c $\frac{1}{2} + \frac{3}{10}$ d $\frac{3}{4} + \frac{3}{12}$

e $\frac{1}{2} + \frac{1}{6}$ f $\frac{1}{3} + \frac{1}{12}$ g $\frac{2}{5} + \frac{3}{10}$ h $\frac{1}{2} + \frac{3}{4}$

i $\frac{3}{5} + \frac{3}{10}$ j $\frac{5}{9} + \frac{1}{3}$ k $\frac{1}{2} + \frac{7}{10}$ l $\frac{2}{3} + \frac{1}{12}$

2 Find the difference between each pair of fractions. Show your working.

a $\frac{3}{4} - \frac{3}{8}$ b $\frac{2}{3} - \frac{1}{6}$ c $\frac{3}{4} - \frac{1}{8}$ d $\frac{1}{3} - \frac{1}{6}$

e $\frac{9}{12} - \frac{1}{2}$ f $\frac{1}{4} - \frac{1}{8}$ g $\frac{9}{10} - \frac{2}{5}$ h $\frac{5}{9} - \frac{1}{3}$

i $\frac{1}{2} - \frac{3}{20}$ j $\frac{9}{20} - \frac{1}{4}$ k $\frac{8}{9} - \frac{2}{3}$ l $\frac{5}{6} - \frac{1}{12}$

Section 3 Number Chapter 6 Fractions

Problem solving

1. What fraction is:

   a $\frac{1}{4}$ more than $\frac{1}{3}$?   b $\frac{1}{2}$ smaller than $\frac{11}{12}$?

2. Billy has $\frac{9}{10}$ of a kilogram of chopped mango. He eats half of it. How much is left?

3. Mr Ali planned his garden so that $\frac{1}{3}$ of the area would be paved, $\frac{1}{6}$ would be for herbs, $\frac{1}{12}$ would be for a small pond and $\frac{1}{4}$ would be planted with lawn. The rest will be flowers and fruit trees. What fraction of the area will be planted with flowers and fruit trees?

4. Make up three challenging word problems involving fractions. Write the problems on a sheet of paper and work out the solutions in your exercise book.

Multiplying fractions and whole numbers

Key maths idea

Ms Smith is planning a pizza party. She estimates that she will need $\frac{1}{2}$ a pizza per person. If she invites five people, how many pizzas will she need?

Ms Smith thinks like this:

> Five lots of one half is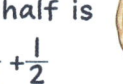
> $\frac{1}{2} + \frac{1}{2} + \frac{1}{2} + \frac{1}{2} + \frac{1}{2}$
> The denominators are the same, so I can add the numerators: $1 + 1 + 1 + 1 + 1 = 5$
> I need 5 halves altogether.
> $\frac{5}{2}$ is the same as $\frac{2}{2} + \frac{2}{2} + \frac{1}{2} = 2\frac{1}{2}$ pizzas

Hint
Remember: any whole number can be written as a fraction with a denominator of 1.

You already know that repeated addition can be written as a multiplication.
$1 + 1 + 1 + 1 + 1 = 5 \times 1 = 5$

When you multiply fractions, you multiply numerators by numerators and denominators by denominators, like this:
$5 \times \frac{1}{2} = \frac{5}{1} \times \frac{1}{2} = \frac{5}{2}$

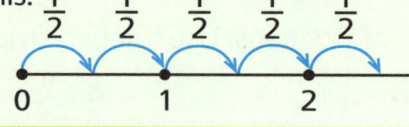

Key word
product

1. Write a multiplication and work out the **product** to answer each question.

   a Amari eats $\frac{1}{3}$ of a bar of chocolate each day of the week. How many bars is that per week?

   b Mariah gives $\frac{3}{4}$ of a cake to each of her four neighbours. How many cakes does she give away?

   c Javid uses $\frac{2}{3}$ of a cup of washing powder per load of washing. How much powder does he use if he does three loads of washing?

60

Multiplying fractions and whole numbers

2 Find the products. Draw diagrams to help.
   a $\frac{1}{2} \times 3$
   b $\frac{1}{4} \times 3$
   c $4 \times \frac{1}{5}$
   d $3 \times \frac{1}{8}$
   e $4 \times \frac{2}{3}$
   f $3 \times \frac{3}{8}$
   g $2 \times \frac{3}{4}$
   h $5 \times \frac{3}{10}$

3 Calculate. If your answer is an improper fraction, convert it to an equivalent mixed number.
   a $5 \times \frac{1}{3}$
   b $\frac{1}{2} \times 7$
   c $\frac{1}{4} \times 5$
   d $\frac{1}{6} \times 5$
   e $\frac{2}{3} \times 3$
   f $\frac{2}{7} \times 4$
   g $\frac{3}{4} \times 6$
   h $\frac{5}{6} \times 3$
   i $\frac{2}{9} \times 4$
   j $\frac{2}{5} \times 7$
   k $\frac{7}{9} \times 2$
   l $6 \times \frac{2}{5}$

Mental maths

1 Use the picture to help you work out:
   a $\frac{1}{2}$ of 12
   b $\frac{1}{3}$ of 12
   c $\frac{1}{12}$ of 12
   d $\frac{3}{4}$ of 12
   e $\frac{1}{4}$ of 12
   f $\frac{1}{6}$ of 12
   g $\frac{5}{6}$ of 12
   h $\frac{7}{12}$ of 12
   i $\frac{1}{2}$ of 24

Hint
$\frac{1}{3}$ of 12 means the same as $\frac{1}{3} \times 12$.
In maths, the word 'of' means multiply.

Problem solving

1 Five friends share 30 sweets equally.
   a How many sweets does each person get?
   b How does the number of sweets change if the sweets are shared between six friends instead of five?

2 There are 12 passengers in a maxi-taxi.
   a One-sixth of them get off. How many passengers are still on the maxi-taxi?
   b At the next stop, $\frac{2}{5}$ of the remaining passengers get off and three more passengers get on. How many passengers are in the maxi-taxi now?

3 Nine friends have seven bars of chocolate to share. Draw a diagram to show how much each friend will get.

4 After a match, 25 soccer players get $\frac{1}{2}$ an orange each. How many oranges are needed for the 25 players? Explain your answer.

5 A large goat has a mass of 42 kilograms. A smaller goat has a mass equal to $\frac{3}{4}$ of the mass of the larger goat. What is the mass of the smaller goat?

Section 3 **Number** Chapter 6 Fractions

Dividing whole numbers by fractions

Key maths idea

Jeevan has five crackers. If he breaks them into halves, how many pieces will he have?

You can model this using a diagram.

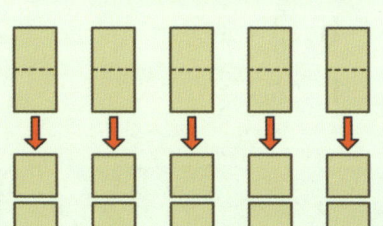

If you divide one whole cracker into halves, you get two parts: $1 \div \frac{1}{2} = 2$

If you divide five crackers into halves, you get ten parts: $5 \div \frac{1}{2} = 10$

$5 \div \frac{1}{2} = 10$ and $5 \times \frac{2}{1} = 10$ $\frac{2}{1}$ is the **reciprocal** of $\frac{1}{2}$

$2 \div \frac{1}{3} = 6$ and $2 \times \frac{3}{1} = 6$ $\frac{3}{1}$ is the reciprocal of $\frac{1}{3}$

To divide by a fraction, multiply by its reciprocal.
But why does this work?

If we are finding $2 \div \frac{1}{3}$, we are asking 'how many $\frac{1}{3}$ pieces fit in 2?'
You can see this on the bar model.

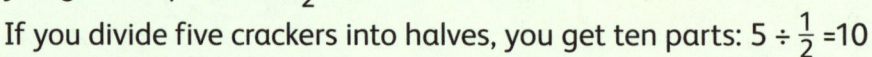

$2 \div \frac{1}{3} = 2 \times \frac{3}{1} = 6$

That is why $2 \div \frac{1}{3}$ is the same as $2 \times 3 = 6$

Key word
reciprocal

Hint
Remember: any whole number can be written as a fraction with a denominator of 1, so $\frac{2}{1} = 2$

1 Nikkita has three cakes.

How many pieces will she have if she divides the cakes into:
  a halves?   **b** quarters?   **c** thirds?   **d** eighths?

2 Calculate:
  a $2 \div \frac{1}{3}$ **b** $3 \div \frac{1}{5}$ **c** $4 \div \frac{1}{8}$ **d** $5 \div \frac{1}{4}$

3 Lisa has 4 metres of ribbon and Malaika has 6 metres of ribbon. They each divide their lengths of ribbon into thirds. How many pieces will they end up with altogether?

Real-life maths

We often divide by fractions without realising it. For example, we might divide our time into half-hour or quarter-hour intervals. Some sports use fractions to divide up the length of a game. For example, a soccer match has two halves, and a basketball game has four quarters.

Dividing whole numbers by fractions

Key maths idea

Now think about $6 \div \frac{2}{3}$

When we divide by $\frac{2}{3}$, we are trying to find out how many groups of two one-thirds we can make.

The reciprocal of $\frac{2}{3}$ is $\frac{3}{2}$ $\qquad 6 \times \frac{3}{2} = \frac{18}{2} = 9$

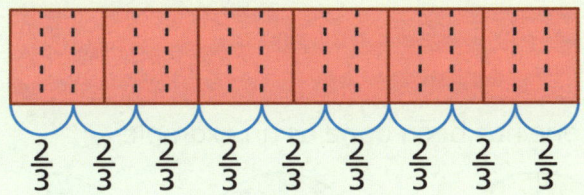

$6 \div \frac{2}{3} = \frac{6}{1} \times \frac{3}{2} = \frac{18}{2} = 9$

9 groups of $\frac{2}{3} = \frac{18}{2}$

1 Divide. Show how you work out each answer.

    a $3 \div \frac{2}{3}$ **b** $6 \div \frac{3}{4}$ **c** $4 \div \frac{2}{5}$ **d** $6 \div \frac{5}{6}$

2 Maleek plans to study for 6 hours in total. He wants to divide the time into $\frac{3}{4}$ hour study sessions. How many study sessions will he have?

3 Jabari divides 12 by $\frac{2}{3}$ and Keshon divides 12 by $\frac{3}{4}$. Without dividing, estimate who will get the greater answer. Do the divisions to check whether you are correct.

Problem solving

1. What is the total mass of 9 bags of chips, if each bag has a mass of $\frac{1}{4}$ kg?
2. Abigail ran $\frac{3}{4}$ kilometre each day for 15 days. How many kilometres is that altogether?
3. Mrs Maraj applies $\frac{2}{5}$ of a litre of plant food to her garden every week for 20 weeks. How many litres will she use in this time?
4. A market garden has 9 barrels of water that they use to water the vegetables. If they use $\frac{2}{3}$ of a barrel each day, how long will the water last?
5. Mr Francis has 3 sacks of cement. If this is $\frac{2}{5}$ of the amount he needs, how much more will he need?

Hint

In a test, you will need to work out which operations you need to do to solve problems. Always read the problem carefully. Draw a diagram to help you visualise the problem and see what you need to do to solve it.

What did you learn?

Look back at the work you did in this chapter. Rate your progress.

1 = I cannot do this. 2 = I need more practice. 3 = I understand it and feel confident.

Can you:
- use equivalent fractions?
- convert between improper fractions and mixed numbers?
- add and subtract fractions?
- multiply with fractions?
- divide a whole number by a fraction?
- solve problems involving fractions?

Section 3 Number Chapter 6 Fractions

Review: Fractions

Key terms and concepts

1 Give an example using numbers to show that you understand each concept.
   a Adding and subtracting fractions with different denominators
   b Calculating a fraction of a group or amount
   c Multiplying fractions by whole numbers
   d Dividing a whole number by a fraction

Quick check

1 Write an equivalent fraction for each model. The first one has been done as an example.

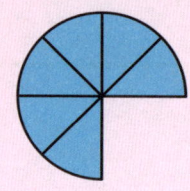 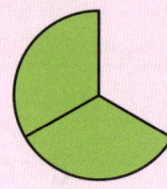  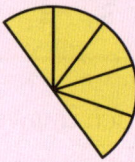 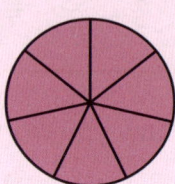 

$\frac{6}{8} = \frac{3}{4}$ $\frac{2}{3} = \square$ $\frac{2}{6} = \square$ $\frac{5}{10} = \square$ $\frac{7}{7} = \square$ $\frac{1}{2} = \square$

2 Do these calculations mentally. Write the answers in your exercise book.
   a $\frac{3}{5} + \frac{1}{5}$ b $1 - \frac{1}{2}$ c $\frac{9}{10} - \frac{3}{10}$ d $1\frac{3}{4} - \frac{1}{4}$
   e $2 - 1\frac{1}{2}$ f $\frac{1}{3} + \frac{1}{3}$ g $2 \times \frac{1}{4}$ h $\frac{1}{2}$ of 16

3 Multiply these numbers.
   a $3 \times \frac{1}{5}$ b $\frac{2}{3} \times 4$ c $5 \times \frac{1}{8}$ d $3 \times \frac{3}{5}$

4 Divide these numbers.
   a $3 \div \frac{1}{2}$ b $4 \div \frac{1}{5}$ c $5 \div \frac{5}{6}$ d $2 \div \frac{2}{7}$

5 One-quarter of the crayons in a container is 9. How many crayons are in the container?

Challenge and investigate

1 Look at the bar models. Make up a word problem to match each one.
   a
   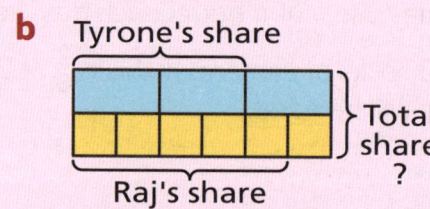

2 Alvin has 5 litres of juice. Omar has $\frac{2}{3}$ of this amount. How much does Omar have?

3 Selina spent $\frac{1}{5}$ of her money on a book. If the book cost $15, how much money did she have to start with?

4 On Maria's market stall, $\frac{1}{4}$ of the space is used for fruit and $\frac{5}{12}$ is used for vegetables. The rest is used for craft items.
   a What fraction of the space has been used?
   b What fraction of the space is available for craft items?

SECTION 3

Chapter 7 Decimals

In this chapter, you will:
- extend your knowledge of place value to include tenths and hundredths
- read, write, order and compare decimals
- round decimals to the nearest whole number and tenth
- add and subtract decimals
- solve problems involving decimals.

Key words
decimals
decimal point

Starting point

Many of the numbers we see around us in daily life are **decimals**. This means they are written with a **decimal point** (a dot). For example, juice is sold in 1.5 litre bottles and the price of gas is $26.87 per gallon.

Look at the decimals in the pictures.

1 Answer these questions with your partner.
   a What is similar about the decimals?
   b How are the decimals different?
   c Where else have you seen numbers written with a decimal point in daily life? Try to give at least three examples.

Real-life maths

Calculators have a decimal-point key so that you can enter measurements, money amounts and other decimals. Find the decimal-point key on your calculator and use it to enter some of the decimals shown in the pictures.

decimal point key

Section 3 Number Chapter 7 Decimals

Tenths

Key words
decimal places
tenths

Key maths idea

Decimals (such as 0.3 or 4.5) allow us to write fractions and mixed numbers using place value. The decimal point separates the whole numbers and the fractional parts. Digits to the right of the decimal point are fractions of a whole. Look at the diagram below. The square represents one whole that has been divided into tenths: $1 = \frac{10}{10}$

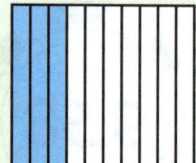

$\frac{3}{10}$ of the square is shaded blue. $\frac{7}{10}$ of the square is not shaded.
The shaded part and the unshaded part of the square can be written as a fraction or as an equivalent decimal.

Three tenths = $\frac{3}{10}$ = 0.3   Seven tenths = $\frac{7}{10}$ = 0.7

The places to the right of the **decimal point** are decimal places. The decimal 0.3 has one decimal place. Both 0.3 and 0.7 have no ones, so we write a 0 in the ones place to the left of the decimal point.

Tenths are smaller than ones, so we extend the place value table to the right to include the new place: **tenths**.

Tens	Ones	.	tenths
	0	.	3
	0	.	7

If you enter 3 ÷ 10 into your calculator, you will see the decimal equivalent 0.3. Your calculator will change any fraction into a decimal, for example $\frac{1}{4}$ = 1 ÷ 4 = 0.25. Try it!

Complete these questions either in pairs or individually.

1 Each large rectangle represents one whole. Write the coloured blocks as a fraction and an equivalent decimal.

  a 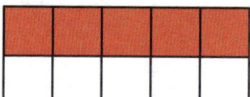   b 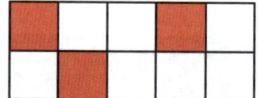   c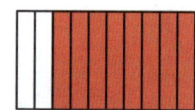

2 Write the equivalent decimal for each fraction.

  a $\frac{1}{10}$ b $\frac{9}{10}$ c $\frac{2}{10}$ d $\frac{7}{10}$ e $\frac{4}{10}$

3 Match each fraction with the correct decimal.

  $\frac{3}{10}$ $\frac{7}{10}$ $\frac{1}{2}$ $\frac{9}{10}$ 0.5   0.7   0.9   0.3

4 Draw a number line like the one below. Fill in all the fractions and decimals.

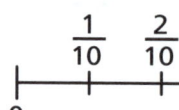

Hint
Use equivalent fractions and your number line to help you.

5 Write the decimal equivalent of:

  a $\frac{1}{2}$ b $\frac{1}{5}$ c $\frac{4}{5}$ d $\frac{2}{10} + \frac{1}{10}$

Hundredths

Key maths idea

The large square represents one whole. The whole has been divided into one hundred small squares. Each small square is one hundredth of the whole.

$1 = \frac{100}{100}$

25 out of the hundred small squares have been shaded.

25 hundredths = $\frac{25}{100}$ = 0.25

The decimal 0.25 has two decimal places. To show this on a place value table, we include another place to the right of the tenths.

Ones	.	tenths	hundredths
0	.	2	5

- The decimal 0.25 has no ones, two tenths and five hundredths
- The 2 is in the tenths place. It has a value of $\frac{2}{10}$ or 2 tenths or two tenths
- The 5 is in the hundredths place. It has a value of $\frac{5}{100}$ or 5 hundredths or five hundredths

1 Write the fraction and the equivalent decimal for the shaded part of each square.

a 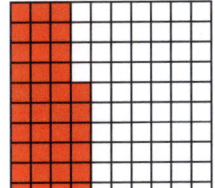   b 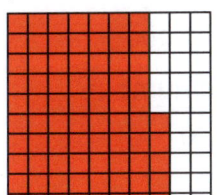   c 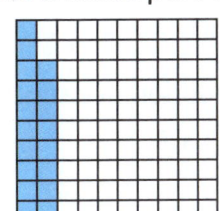   d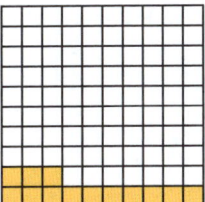

2 Write the equivalent decimal for each fraction.

a $\frac{17}{100}$ b $\frac{33}{100}$ c $\frac{90}{100}$ d $\frac{99}{100}$

e $\frac{30}{100}$ f $\frac{3}{100}$ g $\frac{46}{100}$ h $\frac{51}{100}$

3 Write these decimals as fractions.

a 0.50   b 0.48   c 0.25

d 0.60   e 0.04   f 0.11

4 Which of these decimals are greater than $\frac{1}{2}$?

0.45   0.50   0.55   0.59   0.05

5 Which of these decimals are smaller than 0.45?

0.4   0.5   0.39   0.05   0.19

Section 3 **Number** Chapter 7 Decimals

Key maths idea

Some decimals have whole-number parts, as well as fractional parts. The whole numbers are written to the left of the decimal point.

The number 2.43 is shown on the place value mat.

We can write 2.43 in expanded notation:
2 + 0.4 + 0.03

We can also write 2.43 as a mixed number:
2.43 is equivalent to $2\frac{43}{100}$

O	•	Tths	Hths

1 Write the decimal for each mixed number.

a $1\frac{3}{10}$ b $3\frac{4}{10}$ c $1\frac{12}{100}$ d $2\frac{5}{10}$

e $1\frac{9}{100}$ f $4\frac{99}{100}$ g $2\frac{55}{100}$ h $3\frac{8}{100}$

2 Write each number.

a A number less than 1 with a 2 in the tenths place
b A number with 3 ones, 4 tens and 9 tenths
c A number with 8 tenths, 3 ones, no tens and 2 hundreds

3 Write each expanded notation as a decimal.

a $10 + 2 + \frac{3}{10} + \frac{9}{100}$ b $300 + 8 + \frac{4}{10} + \frac{3}{100}$ c $9 + \frac{3}{100}$

4 What is the value of the blue underlined digit in each number?

a 2<u>3</u>.14 b <u>1</u>0.09 c 3.0<u>7</u> d 4.<u>4</u>5 e 5.9<u>9</u>

Problem solving

1 Look at the digit 9 in each number.

27.9	9.23	0.49	3.91	90.5

a Which number has the greatest value of 9?
b Which two numbers have the same value of 9?
c Which number has 9 in the hundredths place?
d What is the value of the 9 in 0.49?

2 List the decimals which represent the values of the six below:

a six tenths b six hundredths c six tenths and six hundredths

| 65.3 | 9.16 | 0.61 | 16.66 | 6.63 | 66.06 | 3.26 | 1.06 | 66.02 | 3.66 |

3 Find five ways of writing the number 2 as a sum of:

a two one-digit decimals
b two two-digit decimals

68

Money and measurements

Key maths idea

Amounts of money of less than a whole dollar can be written as decimals:
- 1 dollar = 100 cents, so 1 cent is one hundredth of a dollar, or 0.01 dollars
- 25 cents = 25 hundredths of one dollar: $\frac{25}{100}$ = 0.25 dollars

When we work with metric units of length, mass and capacity, we use decimals to show measurements that are less than a whole unit. For example:
- 8 tenths of a kilogram = 0.8 kilograms
- 23 centimetres = $\frac{23}{100}$ metres = 0.23 metres

25 cents is one-quarter of a dollar. That means that 0.25 must be equal to $\frac{1}{4}$.

1 How would you write these amounts of money using decimals?
  - **a** five dollars and seventy-five cents
  - **b** nine dollars and ninety-nine cents
  - **c** twelve dollars and 6 cents
  - **d** one hundred and nineteen dollars and fifty cents

2 Ato has $10.99. The decimal part of this number represents $\frac{99}{100}$ of a dollar, or 99 cents.

What does the decimal part of each of these measurements represent?
  - **a** 4.5 kilograms
  - **b** 1.39 metres
  - **c** 2.45 seconds
  - **d** 89.9 kilometres
  - **e** $8.05
  - **f** 100.45 litres

Full STEAM ahead — Make a decimal place value mat

You are going to make your own decimal place value mat so that you can use it to show and compare decimals and model calculations.

You will need:
- a sheet of thick paper or card
- a ruler
- coloured markers.

Draw a table like this:

Hundred thousands	Ten thousands	Thousands	Hundreds	Tens	Ones	.	tenths	hundredths

Decide how you will model numbers. You could use counters, seeds, cubes or any other small objects. You could also cover your place value mat in plastic and use wipe-off markers to draw dots or shapes to show the numbers.

Section 3 **Number** Chapter 7 Decimals

Compare and order decimals

> **Key maths idea**
>
> We can compare decimals to decide which one is greater or equal. We order decimals by writing them in size order.
>
> You can use place value or number lines to compare and order decimals. You can also model the numbers on your place value mat.
>
> ## Using place value
>
> **Which is greater: 12.38 or 12.83?**
> 1. Stack the numbers.
>   Line up the place values.
>   12.38
>   12.83
> 2. Compare place by place from the left.
>   Find the first place with different digits.
>   12.38
>   12.83
>   ↑ 8 is greater than 3
>
> So 12.83 > 12.38
>
> **Write in ascending order: 4.3, 4.03, 4.33**
> **Step 1**
>   4.30   Add 0 as a placeholder
>   4.03
>   4.33
>
> **Step 2**
>   4.30
>   4.03   1st
>   4.33
>   ↑ 0 is smaller than 3, so 4.03 is the smallest number
>
> **Step 3**
> Now compare the remaining two numbers.
>   4.30   2nd
>   ~~4.03~~
>   4.33   3rd
>   ↑ 0 is smaller than 3, so 4.30 is the next greatest number
>
> **Step 4**
> Write the original numbers in ascending order:
> 4.03, 4.3, 4.33
>
> ## Using a number line
>
> Write these lengths in descending order: 1.3 m, 1.7 m, 1.38 m, 1.5 m, 1.8 m, 1.75 m
> 1. Sketch a number line marked in tenths.
> 2. Write the tenths in the correct position.
> 3. Estimate where the hundredths values will go. Remember that 1.3 is the same as 1.30 and 1.4 is the same as 1.40, so 1.38 will be between those two points.
>
>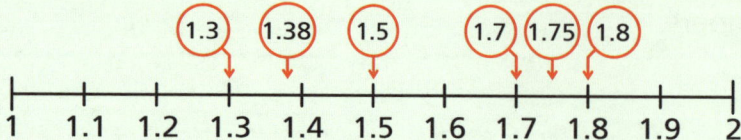
>
> 4. Write the lengths in order from greatest to smallest: 1.8 m, 1.75 m, 1.7 m, 1.5 m, 1.38 m, 1.3 m

Compare and order decimals

Complete these questions either in pairs or individually.
The number line shows the hundredths between 0 and 1.

1   Find these decimals on your number line. Place your ruler on it to show the position.
    a 0.65   b 0.99   c 0.09
    d 0.55   e 0.25   f 0.68

2   Use the number line to find the decimal that is:
    a 0.05 more than 0.6   b 0.05 less than 0.3
    c twelve hundredths more than 0.50   d fifteen hundredths less than 1
    Explain how you found each number.

Hint
Compare your answers with a partner to make sure you have found the correct position.

Mental maths

1 Work out the rule and then write the next two decimals in each counting pattern.
   a 0.68, 0.70, 0.72, …   b 0.45, 0.5, 0.55, …   c 0.96, 0.94, 0.92, …
   d 0.77, 0.67, 0.57, …   e 1.24, 1.28, 1.32, … f 1.87, 1.83, 1.79, …

2 Compare these decimals. Use <, = or >.
   a 0.37 ☐ 0.63   b 0.69 ☐ 0.96   c 0.5 ☐ 0.52   d 0.19 ☐ 0.09
   e 0.21 ☐ 0.42   f 0.45 ☐ 0.4   g 0.11 ☐ 0.13   h 0.49 ☐ 0.38

3 Write each set of decimals in ascending order.
   a 0.58, 0.83, 0.66, 0.49   b 1.25, 1.5, 1.12, 1.15
   c 0.32, 0.23, 0.27, 0.17   d 0.6, 0.55, 0.5, 0.61

4 Write these amounts in order from the most to least amount of money:
   $5.00 $0.55 $5.50 $5.05 $0.05 $5.55

Problem solving

1 Four students made decimals between 0 and 1. Read the clues and write the decimal each student made.
   Priya: My decimal has six hundredths, and the tenths digit is one-third of 12.
   Rishi: My decimal has the same number of hundredths as Priya's but only half the number of tenths.
   Natasha: My decimal has four tenths, and the hundredths digit is the first prime number.
   Reza: My decimal is a multiple of six. It has half as many tenths as hundredths.

2 In a gymnastics competition, five gymnasts get a score out of 10.
   These are the scores of four gymnasts: 9.9   9.81   9.83   9.82
   a What is the lowest score the fifth gymnast in the competition needs to win?
   b If the fifth gymnast wins, which scores are second and third?

Section 3 **Number** Chapter 7 Decimals

Rounding decimals

Key maths idea

You already know how to round whole numbers to a given place.
You can round decimals to whole numbers or to a given decimal place using the same rules.

Round 1.37 m to the nearest tenth

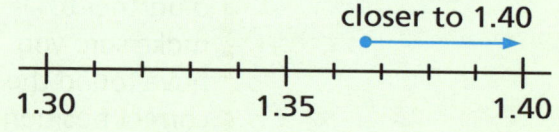

closer to 1.40

1.30   1.35   1.40

↓ This is the tenths place
1.37
↑ The digit to the right is greater than 5, so round up the 3 and get 1.40
We write 0 in any place we round off to show the decimal is approximate

Round 3.56 kg to the nearest kilogram

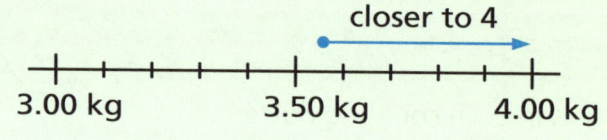

closer to 4

3.00 kg   3.50 kg   4.00 kg

↓ This is the whole kilogram part
3.56
↑ The digit to the right is 5, so round up to 4 kilograms: 4.00 kg

Mental maths

1 Look at these pieces of string.
   a Which one is approximately 1.3 metres long?
   b Tell your partner how you decided.

1.03 m

1.29 m

1.39 m

1 Round each number to the nearest tenth.
   a 0.35   b 0.31   c 0.65   d 1.05
   e 12.33 f 0.99   g 3.21   h 9.17
   i 2.72   j 3.95   k 1.05   l 10.25

2 Round each mass to the nearest kilogram.
   a 3.59 kg   b 2.61 kg   c 3.09 kg
   d 12.32 kg e 19.51 kg f 9.9 kg

3 Round each amount to the nearest whole dollar.
   a $12.25   b $12.50   c $12.99   d $12.48

4 Find five numbers that will round to 3.50 when they are rounded to the nearest tenth. Write your numbers in **ascending** order.

5 Find five lengths that will give you approximately 3 metres when they are rounded to the nearest whole metre. Write your lengths in **descending** order.

72

6 Amar says that 4.82 rounded to the nearest tenth is 4.70 because the 2 in the hundredths place tells you to round down. Explain his mistake and write the correct answer.

7 A number is rounded to the nearest tenth to get 2.50. What is the highest and lowest the original number could be?

8 A taxi travelled 41.6 km on Friday, 121.4 km on Saturday and 59.5 km on Sunday. Round each distance to the nearest kilometre, and then calculate the approximate distance the taxi travelled altogether.

Adding and subtracting decimals

Key maths idea

You can draw number lines and models to help add and subtract decimals.
Read through the examples carefully.

What is 1.7 + 2.5?

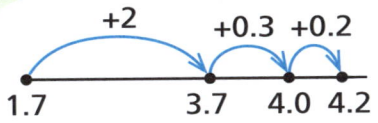

1.7 + 2.5 → 2.5 = 2 + 0.5

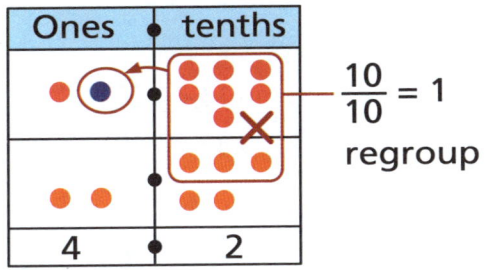

1.7 + 2.5 = 4.2

Calculate 4.25 − 2.13.

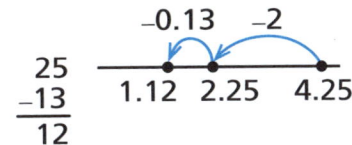

$$\begin{array}{r} 25 \\ -13 \\ \hline 12 \end{array}$$

4.25 − 2.13

O	Tths	Hths
●● ⊗⊗	● ⊗	●⊗ ●⊗ ⊗	−2.13

| 2 | 1 | 2 |

4.25 − 2.13 = 2.12

You can add and subtract decimals in columns using place value in the same way as you add and subtract whole numbers.

When you line up decimals in columns, line up the decimal point, as well as the places.
Fill in empty places using 0 as a placeholder to make it easier to add or subtract in columns.

Add 1.25 and 3.04

Estimate 1 + 3 = 4

1.25 Write the numbers in columns
3.04 Line up the decimal points
―――
4.29 Add the numbers

What is 3.4 + 12.69?

Estimate 3 + 13 = 16

  1
3.40   Write 0 in the empty place
12.69 Line up the decimal points and all the places
―――――
16.09 Add the numbers, regrouping where needed

Section 3 Number Chapter 7 Decimals

(continued)

What is 7.85 − 2.34?
Estimate 8 − 2 = 6

```
  7.85    Write the numbers in columns
  2.34    Line up the decimal points
  ----
  5.51    Subtract the numbers
```

Find the difference between 2 and 0.72
Estimate 2 − 1 = 1

```
       9  1
  2.0̶0̶    Fill the empty places with 0s
  0.72    Line up the decimal points
  ----
  1.28    Subtract, regrouping where needed
```

You can check the answers if you round to the nearest whole number and add or subtract mentally. This helps you decide whether your answer is reasonable.

1 Calculate mentally:
   a 0.3 + 0.4   **b** 1.5 + 0.4   **c** 2.4 + 2.3   **d** 0.5 + 0.4
   e 0.9 − 0.2   **f** 1.5 − 1.2   **g** 0.8 − 0.8   **h** 1.8 − 0.5
   i 0.8 − 0.3   **j** 12.3 + 2.2 **k** 10.5 − 3.5 **l** 6.5 + 1.5

2 Add:
   a 0.61 + 0.25   **b** 0.43 + 0.26   **c** 0.15 + 0.83   **d** 4.12 + 4.27
   e 3.26 + 2.45   **f** 0.83 + 3.21   **g** 3.08 + 2.8   **h** 9.85 + 2.19

3 Subtract:
   a 6.7 − 2.6   **b** 0.98 − 0.23   **c** 0.99 − 0.38   **d** 4.37 − 1.06
   e 3.45 − 1.07 **f** 4.2 − 2.18   **g** 3.08 − 2.1   **h** 8.5 − 4.41

4 Round each decimal to the nearest whole number and estimate the answer before doing each of these calculations. Take note of the operation signs.
   a 3.63 + 8.7 + 4.31   **b** 12.09 + 3.9 + 2.8   **c** 45.05 + 23.97   **d** 12.23 + 6.99
   e 16.08 − 9.06   **f** 12.50 − 7.03   **g** 19.97 − 4.58   **h** 14.23 − 7.05

5 Ms Roberts asked her class to add 0.24, 12.4 and 3.8. Who set their work out correctly? Explain what the others did wrong.

```
Derrick          Oneika           Chin             Patricia
  0.24             0.24             024              0.24
 12.4             12.4             124             12.40
 +3.8            + 3.8            +38             + 3.80
```

Talking maths

Jayson added the following on his calculator: 13.4, 0.56, 135.2 and 20.6.
He correctly estimated the answer to be 170 but his calculator showed 48.08.
What do you think Jayson did wrong?

Rounding decimals

Key maths idea

Draw **bar models** to help you visualise problems and see how to solve them.
Read through the examples to see how to do this.

Key word
bar model

Example 1
Jaden is 1.52 m tall. Keshon is 0.12 m taller than Jaden. How tall is Keshon?

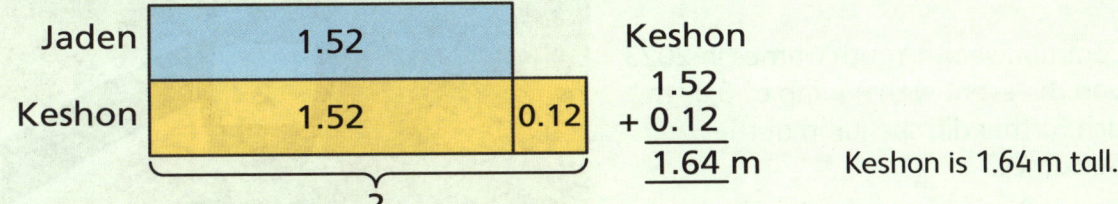

Keshon
$$\begin{array}{r} 1.52 \\ +\,0.12 \\ \hline 1.64 \text{ m} \end{array}$$

Keshon is 1.64 m tall.

Example 2
Sharon has 2.6 kg of sugar. She uses 0.8 kg. How much does she have left?

$$\begin{array}{r} {}^1\!\!\not{2}.{}^1\!6 \\ -\,0.8 \\ \hline 1.8 \text{ kg} \end{array}$$

Sharon has 1.8 kg left.

Example 3
Jemila has a 1.2 m length of ribbon. Mariah has a length of ribbon 0.45 m longer than Jemila's piece. What is the combined length of the ribbons?

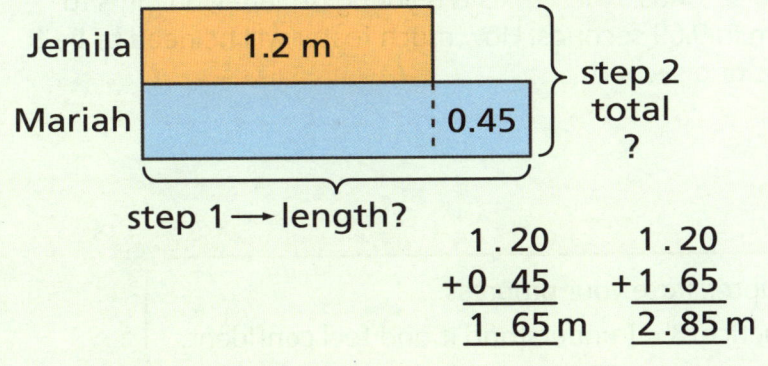

$$\begin{array}{r} 1.20 \\ +\,0.45 \\ \hline 1.65 \text{ m} \end{array} \qquad \begin{array}{r} 1.20 \\ +\,1.65 \\ \hline 2.85 \text{ m} \end{array}$$

Mariah's ribbon is 1.65 m long, so the combined length is 2.85 m.

> **Hint**
> The bar model is not the solution; it just helps you to see what you need to do to work out the solution. Use the working-out space on the test paper to draw bar models to show your working and thinking.

Section 3 **Number** Chapter 7 Decimals

Problem solving

1. Janae De Gannes came first in the long jump event at the CARIFTA Games in 2024. She jumped 6.5 m, breaking the record of 6.48 m set by Yanis David at the 2016 games.

   a. How much further did Janae jump than Yanis?

   b. At the Commonwealth Youth Games in 2023, Janae won the event with a jump of 6.07 m. How much further did she jump at the CARIFTA Games?

   c. In January 2024, Janae jumped 6.25 m at the NAAATT event. How much shorter was her 2023 jump?

   d. If Janae beats her own 6.5 m record by 12 cm, what will her new jump record be?

2. The world-record times for the women's 100 m and 200 m sprints are both held by Florence Griffith Joyner. The 100 m record is 10.49 seconds and the 200 m record is 21.34 seconds.

   a. In 2021, Elaine Thompson-Herah of Jamaica ran the 200 m in 21.53 seconds. How much faster would she need to run to beat the record?

   b. In 2023, Shericka Jackson missed the 200 m record by 0.07 seconds. What was her time in that race?

   c. When Florence Griffith Joyner set the 100 m record, she beat the previous record by 0.27 seconds. What was the previous record?

3. Usain Bolt set the 100 m record of 9.58 seconds. Noah Lyles is a young athlete who aims to beat that record. In 2024, he ran 100 m in 9.69 seconds. How much faster will he need to be to beat Usain Bolt's time by 3 hundredths of a second?

What did you learn?

Look back at the work you did in this chapter. Rate your progress.
1 = I cannot do this. **2** = I need more practice. **3** = I understand it and feel confident.

Can you:
- use place value tables to include tenths and hundredths?
- read, write, order and compare decimals?
- round decimals to the nearest whole number and tenth?
- add and subtract decimals?
- solve problems involving decimals?

Review: Decimals

Key terms and concepts

1 Complete the sentences to summarise what you learned.
   a A decimal is written with a ____ to separate the ____ and the ____ parts of the number.
   b Numbers to the left of the decimal point are ____.
   c Numbers to the right of the decimal point are ____.
   d You use ____ to compare the size of decimals. We compare digits from the ____.
   e When you add or subtract decimals, it is important to ____.

Quick check

1 What is the value of the 5 in each number?
   a 2.5   **b** 0.25   **c** 5.89   **d** 52.07   **e** 12.58

2 Write each fraction as an equivalent decimal.
   a $\frac{23}{100}$ **b** $\frac{6}{10}$ **c** $\frac{1}{2}$ **d** $\frac{9}{100}$ **e** $1\frac{3}{10}$

3 Write each set of measurements in **ascending** order.
   a 2.04 kg   2.41 kg   2.14 kg   2.4 kg
   b 1.3 ℓ   0.13 ℓ   13.01 ℓ   0.31 ℓ

4 Round each value to the nearest tenth.
   a 12.09   **b** 3.42   **c** 9.53   **d** 4.55   **e** 1.98

5 Round each value to the nearest whole number and estimate the answer before doing each calculation. Use your estimate to check your answer is reasonable.
   a 12.35 + 15.67   **b** 132.87 + 12.6   **c** 12.08 + 18.4   **d** 12.87 + 3.45
   e 34.76 – 12.1   **f** 14.8 – 12.35   **g** 42 – 13.51   **h** 9.08 – 1.99

Challenge and investigate

1 Round each decimal to the nearest whole number.
   a 16.4   **b** 28.7   **c** 0.99   **d** 3.83   **e** 69.09

2 List the next four decimals in each pattern.
   a 2.51, 2.50, 2.49, …   **b** 5.05, 5.1, 5.15, …

3 Ms Persad asks the class to write twelve dollars and seventy-five cents using numbers.
   a Which of these students has written the amount correctly?

Maleek	Diane	Marcus	Abigail	Tyrone
$1275	$127.5	$12.75	$1275.00	$1.275

   b What mistakes have the others made?
   c How would you write 15 dollars as a decimal?
   d What is your height in metres and centimetres? Write it as a decimal.

4 Calculate. Show your working.
   a 23.47 + 38.32   **b** 12.09 + 3.7   **c** 123.09 – 19.02   **d** 14.5 – 6.38

5 Amari has 2 litres of water. He pours out 1.05 litres. How much is left?

6 Sasha has a laptop and a printer next to each other on her desk. The desk is 1.2 m wide. The laptop is 32.5 cm wide and the printer is 0.48 m wide. Is the desk wide enough for her to put a 0.5 m-long drawer unit next to them?

SECTION 4

Chapter 8 Measuring length, mass and time

In this chapter, you will:
- learn about standard units of measurement
- use mm, cm and km in questions about length
- use g and kg to measure mass
- use minutes and hours to solve problems about time
- solve a variety of questions involving measurement.

Starting point

1. Look at these items. Which of them are used to measure:
   a time?
   b length?
   c mass?

2. a Which of these items do you use most often? What do you use it for?
   b Which of the items do you use least often or never use?
   c Choose three of the items. Think of a job or a career where that item might be important to have. Say why.

3. For each item, say what the numbers on it mean.

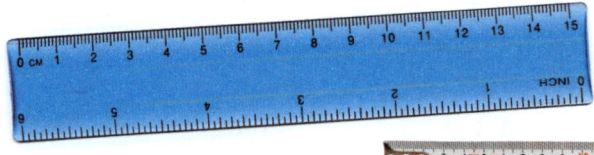

Talking maths

Have you ever heard these expressions? What do they mean? Which units of measurement do they use to express their meaning?

- I am getting there inch by inch.
- He is worth his weight in gold.
- The children ate a ton of candy at the party.
- They went the extra mile.

Metres and centimetres

Key maths idea

When we ask 'How long is it?', we are talking about **length**. We use rulers and tape measures to measure length. The markings on a ruler show you the units.

The standard **unit** for measuring length is the **metre** (m).

1 metre is made up of 100 centimetres (cm).

A child's armspan is about 1 metre long.

Key words
length
unit
metre

Real-life maths

1 Do this activity in your classroom or at home.
   a Identify three objects that are shorter than 1 metre, three that are about 1 metre and three that are longer than 1 metre.
   b Estimate the length of each object in metres or centimetres.
   c Use a tape measure or ruler to measure the objects.
   d Record your findings in a table like this in your exercise book.

Shorter than 1 metre	About 1 metre	Longer than 1 metre

Mental maths

1 Write each decimal as a fraction of 100, then write it in its simplest form.
   a 0.1   b 0.01   c 0.5
   d 0.05   e 0.2   f 0.02

1 How many centimetres are there in:
   a half a metre?   b one quarter of a metre?   c one tenth of a metre?

2 Express each length in metres. Use decimals.
   a 100 cm   b 200 cm   c 700 cm   d 1000 cm
   e 150 cm   f 425 cm   g 675 cm   h 580 cm

3 100 cm = 1 m. Use this relationship to solve these problems.
   a Dinesh has some pieces of spaghetti that are each 20 cm long. How many must he lay end to end to make a length of 1 m?
   b Sharon has a 1 metre strip of wire. She wants to cut it into strips of 25 cm. How many can she make?
   c A skipping rope is 50 cm long. Amar lays five skipping ropes end to end. What is their total length?

Section 4 Measurement Chapter 8 Measuring length, mass and time

More units of length

Key maths idea

Many things are smaller than a centimetre. Think about the length of an ant, the width of your pencil or the thickness of a front door key. How would you measure those?

To measure shorter lengths, we need a smaller unit.

If you look at your ruler, you can see that 1 cm is divided into ten smaller units. These are **millimetres** (mm).

1 cm = 10 millimetres (mm)

Other distances are much too great to measure in mm, cm or m. Think about the distance between two cities, or between two islands. For long distances, we measure in **kilometres** (km).

1 km = 1000 m

A millimetre is about the same as the thickness of a coin.

Key words
millimetre
kilometre

1. Which unit would you use for the following measurements: mm, cm, m or km? Tell a partner and explain why.
   a. How much you grew in the last year
   b. The thickness of an eraser
   c. The height of your school chair
   d. The length of your foot
   e. The distance from Port of Spain to San Fernando
   f. The distance from one end of the school playground to the other
   g. The thickness of a mobile phone

2. Why do we need a unit smaller than centimetres? Write the reason in your own words, giving some examples of your own.

3. On this ruler, you can see that one side is marked in cm and the other side is marked in mm.

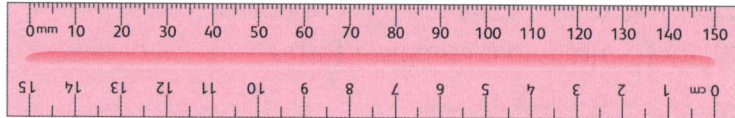

   a. What is similar about the markings on each side, and what is different?
   b. What do you think the shorter markings between the numbers on each side mean?

Talking maths

The words centimetre, millimetre and kilometre are all made from the word 'metre' with a prefix (a small part at the beginning of a word). Here are some other measurement words with the same prefixes:

centilitre   millisecond   milligram   kilogram   kilolitre

1. Talk about what you think these words mean, and what they might be used to measure. You can look them up in a dictionary.
2. How do the prefixes help you to remember the size of each unit?

Estimating and measuring lengths

> **Key maths idea**
>
> At school, we usually use rulers to measure the lengths of objects and lines.
> Make sure the ruler lies straight along the length of the object.
>
>
>
> Make sure you line up the 0 mark on the ruler with the beginning of the object you are measuring.
>
> Before you measure, make an estimate. After you measure, think about whether the answer looks reasonable compared to your estimate.

Use a ruler that is marked in cm and mm.

1. Draw lines of the following lengths.
   - a 15 mm
   - b 22 mm
   - c 37 mm
   - d 5 mm

2. Write the lengths from question **1** in cm. Use decimals.

3. Draw lines of the following lengths.
   - a 6.8 cm
   - b 4.3 cm
   - c 7.5 cm
   - d 0.8 cm

4. Write the lengths from question **3** in mm.

5. Find an item to match each length below. Write what you find. Measure it with a ruler and write the exact measurement in mm.
   - a Less than 1 cm
   - b Between 2 cm and 4 cm
   - c Between 7 cm and 10 cm

Hint
You can measure the length or width (thickness) of an object.

Full STEAM ahead How far can you throw it?

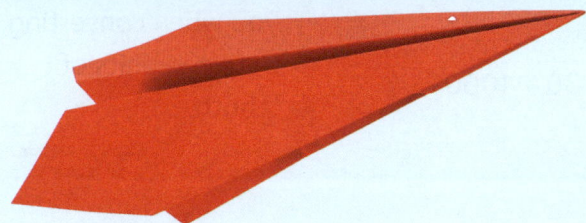

You will need:
- a paper plane or frisbee
- chalk or stones to make marks
- a measuring tape or trundle wheel.

- Use a paper plane or frisbee.
- Go outside.
- Mark a starting point where everyone will stand, using chalk or stones.
- Take turns to throw the object.
- Measure each throw in metres and centimetres to see how far it lands from the starting point.

Section 4 **Measurement** Chapter 8 Measuring length, mass and time

Converting between units of length

Key maths idea

When you solve problems involving length, you sometimes need to **convert** from one unit to another.

1 m = 100 cm   1 m = 1000 mm
1 cm = 10 mm
1 km = 1000 m   1 km = 100 000 cm

Key word
convert

Example
Jenny is planting spinach. The rows must be 20 cm apart.
Jenny's vegetable patch is 1.6 m wide. How many rows can she fit?

Working out	Thinking and reasoning
1.6 m ÷ 20 cm = ?	We cannot work with two different units. We need to convert both to m, or both to cm. Working with cm is easier in this case, as we will work with whole numbers.
1.6 m = ____ cm	First estimate. 1 m = 100 cm, so 1.6 m is more than 100 cm.
1.6 × 100 = 160	There are 100 cm in 1 m.
	To convert from m to cm, multiply by 100.
1.6 m = 160 cm	Check against the estimate. It looks reasonable.
160 cm ÷ 20 cm = 8	Jenny can fit 8 rows of spinach in her vegetable patch.

Mental maths

1 Work with a partner. Take turns doing each set of calculations.

a
20 × 10   200 × 100   200 ÷ 10
200 ÷ 100   2000 ÷ 100

b
1.5 × 10   150 × 100   1.5 ÷ 10
15 × 100   1500 ÷ 100

c
8.7 × 100   870 × 1000   870 ÷ 1000
87 ÷ 100   8.7 ÷ 10

d
0.003 × 100   0.3 × 1000   30 ÷ 1000
0.03 ÷ 100   0.3 ÷ 10

Hint
Practising multiplication and division by 10, 100 and 1000 can help you when converting between units of length.

1 Convert these measurements from m to cm. The first one has been done for you.
  a 2 m = 200 cm   b 3 m = ___ cm   c 5 m = ___ cm
  d 10 m = ___ cm   e 2.5 m = ___ cm   f 6.5 m = ___ cm

2 Use your answers to question **1** to write a rule for converting from m to cm.

3 Convert these measurements from cm to m.
  a 100 cm = ___ m   b 150 cm = ___ m   c 300 cm = ___ m
  d 380 cm = ___ m   e 1500 cm = ___ m f 185 000 cm = ___ m

Measuring mass

4 Use your answers to question 3 to write a rule for converting from cm to m.
5 There are 1000 m in 1 kilometre. Use this relationship to write a rule for:
   a converting from m to km
   b converting from km to m.
6 Meela says that the relationship between mm and m is the same as the relationship between m and km. What does she mean?
7 Convert these measurements.
   a 50 mm = ___ cm
   b 500 mm = ___ cm
   c 500 mm = ___ m
   d 25 mm = ___ cm
   e 125 mm = ___ cm
   f 3125 mm = ___ cm
   g 3125 mm = ___ m
   h $\frac{1}{2}$ m = ___ mm
   i 0.75 m = ___ mm
   j 0.3 m = ___ mm
   k 0.05 m = ___ mm
   l 0.008 m = ___ mm

Problem solving

1 Ria says that 90 m is the same as 9000 mm. What is her mistake?
2 Lee says that $\frac{1}{2}$ metre = 200 mm. What is his mistake?

Measuring mass

Talking maths

1 Discuss with a partner:
   a What is being measured in both of these pictures?
   b What units do you think they are using?
   c What other kinds of scales have you seen in real life?
2 What other ways do we use the word 'scale'?

Key maths idea

When we want to find out how heavy an object is, we measure its **mass**.

The metric units for measuring mass are **kilograms** and **grams**.

1 kilogram (kg) = 1000 grams (g)

You can use this relationship to help you convert between units.

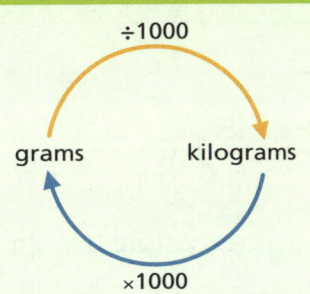

Key words
mass
kilograms
grams

Mental maths

1 How many grams in:
   a 1 kg?
   b 2 kg?
   c half of a kilogram?
2 How many kilograms in:
   a 1000 g?
   b 2000 g?
   c 8000 g?
   d 12 000 g?

Section 4 Measurement Chapter 8 Measuring length, mass and time

1. Do the following items weigh about a gram or about a kilogram?
   - a the cap of a pen
   - b one piece of candy
   - c a bag of sugar
   - d a small laptop
   - e a large feather
   - f a 1-litre bottle of water

2. Would you measure the following items in grams or kilograms? Say why.
   - a a single banana
   - b a large bunch of bananas
   - c a cat
   - d a mobile phone
   - e a paper clip
   - f a watermelon

3. Write each of the following amounts in grams.
   - a 5 kg
   - b 5.5 kg
   - c 0.25 kg
   - d 0.75 kg
   - e 1.8 kg
   - f 3.25 kg
   - g 7.7 kg
   - h 10.1 kg

4. Use your answers to question 3. Write the rule for converting from kilograms to grams.

5. Write each amount in kg.
   - a 5000 g
   - b 500 g
   - c 250 g
   - d 8750 g

6. Use your answers to question 5. Write the rule for converting from grams to kilograms.

Full STEAM ahead Estimate and measure mass

1. Work in a group.
   - a Discuss the objects you will weigh. Guess which will be the heaviest and which will be the lightest.
   - b Arrange them in (estimated) order, from lightest to heaviest. Estimate the weights in grams or kilograms.
   - c Weigh the object you predicted would be heaviest, and the object you predicted would be lightest. Note their masses in grams or kilograms. How accurate were your measurements? Discuss whether you want to change any of your other estimates.
   - d Weigh the rest of the objects. Make a table like this to describe the objects you weighed.

You will need:
- a kitchen scale or balance scale
- a variety of objects for weighing.

Object	Estimate	Actual weight

Problem solving

1. Two customers at a grocery store bought the following items. Calculate the total mass of each customer's items.
   - a Brenda bought 1.2 kg potatoes, 500 g onions and 750 g tomatoes.
   - b Gary bought 325 g grapes, 450 g plums and 1.35 kg peaches.

2. A baker has a 1.5-kg bag of flour. He uses 200 g for a batch of cookies, and 650 g for a loaf of bread. How much flour does he have left?

3. A hiker is packing her backpack. The empty backpack weighs 1.2 kg, and she has 3.5 kg of clothing, 4.6 kg of food and 2 kg 600 g of other supplies. What is the total mass of the pack in kg and g?

Analog clocks

Key maths idea

We tell the time in **hours** and **minutes**.
1 hour = 60 minutes $\frac{1}{2}$ hour = 30 minutes $\frac{1}{4}$ hour = 15 minutes
1 minute = 60 seconds   1 day = 24 hours

Key words
hour
minute
analog clock
hour hand
minute hand

We use watches and clocks to tell the time.
An **analog clock** has hands that turn around the face of the clock.
The short hand is called the **hour hand**. The long hand is the **minute hand**.
When the long hand points to the 12, the time is on the hour. As the long hand moves, it tells us how many minutes have passed since the hour, or how many minutes until the next hour.

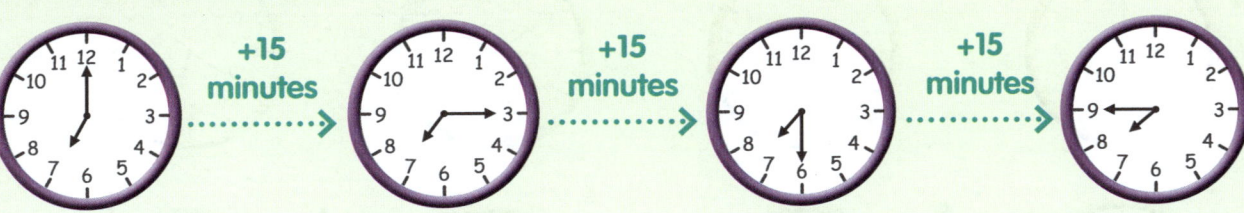

7 o'clock   quarter past 7   half past 7   quarter to 8

1 How long does it take for:
   a the minute hand to make a full turn?
   b the hour hand to make a full turn?
   c the minute hand to make a half-turn?
   d the hour hand to make a quarter-turn?
   e the minute hand to move from pointing directly at the 12 to pointing directly at the 1?
   f the hour hand to move from pointing directly at the 12 to pointing directly at the 1?

2 Use the blank clock face. Cut two strips of card – a long one for the minute hand and a shorter one for the hour hand. Work with a partner. Take turns to show these times on the clock face.
   a 3 o'clock
   b 7 o'clock
   c half past 2
   d half past 10
   e quarter to 11
   f quarter past 5
   g quarter past 1
   h quarter past 9

3 Selina says: 'When the long hand points to the 10, it is ten minutes to the hour, so when the long hand points to the 11, it must be eleven minutes to the hour.' What is her mistake?

Section 4 Measurement Chapter 8 Measuring length, mass and time

> ### Key maths idea
>
> When the long hand is on the 12, the time is on the hour (we say o'clock). For the next half-hour, we count the time in minutes past the hour. After half past, we count minutes until the next hour.
>
>
>
> 5 minutes past 3   25 minutes past 3   20 minutes to 4   10 minutes to 4

1 What time is shown on each clock? Write in your exercise book.

a    b    c    d

e    f    g    h

2 Use the blank clock face and strips of card you used for question **2** on the previous page. Work with a partner. Take turns to show these times.

  a 5 minutes past 3
  b 10 minutes past 8
  c 20 minutes past 7
  d 25 minutes to 9
  e 20 minutes to 12
  f 5 minutes to 1

Problem solving

Answer the questions about each time shown on the clocks below.

1
  a What time will this clock show after 15 minutes?
  b What time was it 20 minutes earlier?
  c How long will it take to get to 3 o'clock?

2
  a How long is it from this time until 8 o'clock?
  b What time will it be half an hour after the time shown on this clock?
  c How much time has passed since half past 6?

Time on the digital clock

Key maths idea

Digital clocks show the time using two numbers separated by the : symbol, which is called a **colon**. The number before the colon gives the hour, and the number after the colon gives the minutes after the hour. Each day begins at midnight, which is 00:00 hours.

Key words
digital clocks
colon

This clock shows 1 minute before 6 o'clock in the morning.

Some digital clocks use the 24-hour time system. With the 24-hour system, the hours of the day are numbered from 0 to 24. Midnight is 00:00, midday is 12:00 and one o'clock in the afternoon is 13:00.

Here are some examples of time expressed in different ways to show you how the 24-hour clock works: After 12 o'clock (midday), the hour times continue to 13:00 (one o'clock), 14:00 (2 o'clock), and so on, up to 23:00 (11 o'clock).

5 minutes earlier

5 minutes to midnight
11:55 p.m.
23:55

midnight
12:00 a.m.
00:00

5 minutes later

5 past midnight
12:05 a.m.
00:05

$1\frac{1}{2}$ hours earlier

half past 10
10:30 a.m.
10:30

noon (midday)
12:00 p.m.
12:00

$1\frac{1}{2}$ hours later

half past 1
1:30 p.m.
13:30

Talking maths

Work in pairs. Take turns. One partner says a time of the day, in words. The other partner writes the time in a different way (24-hour notation, or a.m. and p.m. time). Instead of saying the time of day with words, you can also act it out using your arms as hands of the clock.

Section 4 Measurement Chapter 8 Measuring length, mass and time

1. Write each time in 24-hour notation.
   a. one o'clock in the afternoon
   b. three o'clock in the afternoon
   c. ten o'clock at night
   d. seven o'clock in the evening
   e. 5.00 p.m.
   f. 8.00 p.m.
   g. 2.00 p.m.

2. Explain where the hands would point on an analog clock at these times.
   a. 00:15   b. 02:45
   c. 14:40   d. 15:30
   e. 19:25   f. 20:00
   g. 21:21   h. 22:35

3. Write these times in words. Include the time of day – morning, afternoon or night, for example 20:30 is half past 8 at night.
   a. 23:17   b. 10:42
   c. 21:56   d. 03:38
   e. 05:12   f. 19:31
   g. 18:07   h. 16:34

4. Write each time in 24-hour notation.
   a. midday   b. quarter past 12 in the afternoon
   c. 1 o'clock in the morning   d. 1 o'clock in the afternoon
   e. half past 3 in the morning   f. quarter to 2 in the afternoon

5. Work with a partner. For each number line, work out:
   - how has the number line been divided up?
   - how much time passes from one mark to the next?
   - what should the missing marks say?

   a

   b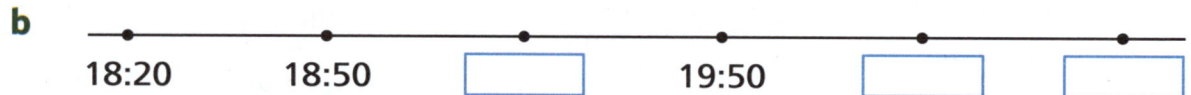

   c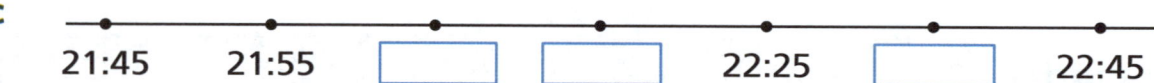

   d

Duration

Key maths idea

The **duration** of an activity or event means how long it takes.
- If you know when something starts and when it ends, you can work out the duration.
- If you know when it starts, and the duration, you can work out the finish time.
- If you know the duration and the finish time, you can work out the start time.

Key words
duration
interval
elapsed time

Counting in time **intervals** can help you work out how long something takes.

For example:

On Monday, my maths class starts at 11:40 and ends at 12:10. What is the duration of the lesson?

I count on in ten-minute intervals. 3 × 10 = 30

The lesson takes 30 minutes or half an hour.

Elapsed time is how much time has passed from one given time to another. Each interval shows an amount of elapsed time. After 10 minutes has elapsed, the time is 11:50. After 20 minutes has elapsed, the time is 12:00. After 30 minutes, the time is 12:10.

Mental maths

1. How many minutes in:
   - **a** half an hour?
   - **b** a quarter-hour?
   - **c** an hour and a half?
   - **d** $\frac{3}{4}$ of an hour?
   - **e** 4 and a half hours?
   - **f** 8 hours?

2. An hour has:
   - **a** ____ 10-minute intervals
   - **b** ____ 20-minute intervals

1. Do the following activities take more than, less than or about an hour?

   (cutting an apple)   (eating breakfast)   (watching a movie)   (a birthday party)

   (travel to a different island)   (breaktime at school)   (cooking an egg)

2. About how long do these activities take? Estimate the duration in hours or minutes.

a
a night's sleep

b
a game of cricket

c
a bus ride to school

d
a beach outing

Section 4 Measurement   Chapter 8   Measuring length, mass and time

3 Here are the starting and finishing times of some different activities. Work out how long each one takes in hours and minutes

	Start time	Finish time
a	4:00 a.m.	7.30 a.m.
c	3:00 a.m.	5:30 p.m.
e	6:19 p.m.	8:19 p.m.

	Start time	Finish time
b	10:05 a.m.	11:25 p.m.
d	12:00 p.m.	2:00 a.m.
f	7:30 a.m.	10:15 a.m.

4 What is the time:
   a   half an hour after 14:20?
   b   2 hours before 01:30?
   c   45 minutes before 18:25?
   d   an hour and a half after 7 o'clock?
   e   4 hours and 20 minutes after 23:55?

Problem solving

1 Adara goes for a run. She leaves the house at 08:15 and gets back at 08:55. How long did the run last?

2 Vanessa goes for a hike. She starts the route at 09:30. She reaches the lookout point at 11:45. She takes a 20-minute break, then takes $1\frac{1}{2}$ hours to walk back.
   a   How long does it take to get from the start to the lookout?
   b   At what time does she finish her break?
   c   What time does she get back?

3 Jeevan's school starts at 07:45 and ends at 13:00.
   a   How much time does he spend at school every day?
   b   How many hours does he spend at school every week?

4 Ashanti goes to a different school. Her school day starts at 07:30 and ends at 15:45. How many hours does Ashanti spend at school:
   a   each day?   b   each week?

5 How much more time does Ashanti spend at school every day than Jeevan?

6 Mrs Clarke travels by bus to work. It takes her 45 minutes to get ready for work. The bus leaves at 07:30. It takes her 10 minutes to walk to the bus stop. At what time must she get up in the morning?

Full STEAM ahead   Holding a plank

1 How long can you hold a plank position?

   Work in small groups.

   Use a watch, clock or mobile phone to measure how long each person can hold a plank position.

   You can also time the duration of other activities, such as:
   - skipping
   - standing on one leg
   - humming a note before you need to take another breath.

Draw a table. List the names of the people in your group. Write the estimated and measured time of your chosen activity. Look back at your estimated time. Was it reasonable?

Calendars and schedules

Key maths idea

A **year** has 12 **months**. April, June, September and November each have 30 days. All the other months have 31 days, except February. February has 28 days in most years, but in a leap year February has 29 days.

7 days in a **week**   about 4 weeks in a month

about 30 days in a month

365 days in a year   366 days in a **leap year** (every fourth year)

A **calendar**, **schedule** or **timetable** helps you to plan your time. Examples of these planners include lesson timetables for school, programme listings for TV and bus and train schedules.

Key words
- year
- month
- week
- leap year
- calendar
- schedule
- timetable

Use the calendar to help you answer the questions.

			March			
Sunday	Monday	Tuesday	Wednesday	Thursday	Friday	Saturday
	1	2	3	4	5	6 family picnic
7	8	9	10	11	12	13 Zaida's birthday
14	15	16	17	18 maths test	19	20
21	22	23	24	25	26	27
28	29 school outing	30	31			

1 What is the date of:
   a Zaida's birthday?   **b** the school outing?   **c** the maths test?

2 On which day of the week will the following dates fall in the year shown in this calendar?
   a the last day of February   **b** the second day of April

3 Billy's birthday is exactly two weeks after Zaida's. What is the date of Billy's birthday?

4 Alvin's birthday is one week after the school outing. What is the date of his birthday?

5 Zaida says she will revise her maths every day from after her birthday until the day of the maths test. How many days does she plan to study?

Section 4 Measurement Chapter 8 Measuring length, mass and time

Mental maths

1. A schedule is marked in 2-hour intervals, starting from 00:15 on one day to the same time the next day. Say the times.
2. In a time-lapse video, the camera takes a photograph every 45 minutes for 12 hours, starting at 16:00. Say the times of the first ten photographs.
3. At a cinema, a film begins every 90 minutes, starting at 10.00 a.m. Say the times of the first eight films of the day.

Problem solving

The picture shows a departures board at an international airport. Use the information in the picture to help you answer the questions.

TIME	DESTINATION	FLIGHT	GATE	REMARKS
18:08	NEW YORK	AC 103	13	CANCELLED
18:16	BERLIN	CI5723	22	CANCELLED
18:38	LONDON	MU5984	12	CANCELLED
18:49	TOKYO	JL 608	14	DELAYED
19:07	HONG KONG	CX6471	25	CANCELLED
19:18	MADRID	IB3941	03	DELAYED
19:29	SYDNEY	LH5021	17	CANCELLED
19:35	TORONTO	KA 197	11	CANCELLED
19:44	PARIS	AF5870	02	DELAYED
19:50	ROME	FM 324	04	CANCELLED

DEPARTURES 18:02

1. What does the time in the top right-hand corner of the board tell us?
2. How much time does each of the following passengers have to wait for their plane to depart?
   a A passenger flying to New York
   b A passenger flying to Berlin
3. How much later is the flight to Toronto than the flight to London?
4. The delayed flight to Paris eventually leaves at 23:15. How much later is this than its scheduled departure?
5. Make up two questions of your own about the times on the departures board. Exchange them with a partner and answer each other's questions.
6. Discuss with your partner:
   a Why do you think flight schedules, bus schedules and other travel timetables usually use 24-hour notation?
   b Where else do you usually see time expressed in this way?

What did you learn?

Look back at the work you did in this chapter. Rate your progress.
1 = I cannot do this. **2** = I need more practice. **3** = I understand it and feel confident.

Can you:
- identify the correct units for measuring length, mass and time?
- measure length in mm, cm and km?
- measure mass in g and kg?
- express durations in hours and minutes?
- solve problems involving the measurement of length, mass and time?

Review: Measuring length, mass and time

Key terms and concepts
Copy and complete.
1. The standard unit of length is the ____. It is made up of 100 ____. The smallest unit of length on a normal ruler is the ____. Ten of these is equal to ____.
2. A cabbage or a pineapple weighs approximately one ____. This is equal to one thousand ____.
3. There are ____ days in a week, around ____ days in a month and usually ____ days in a year.

Quick check
1. Which units would you probably use when you use:
   - **a** a tape measure?
   - **b** a calendar?
   - **c** a kitchen scale?
   - **d** a bathroom scale?
   - **e** a watch?
   - **f** a ruler?
2. Draw lines that measure:
   - **a** 15 mm
   - **b** 3.5 cm
   - **c** 78 mm
3. Randy's height is 1 m 48 cm. Write that in cm.
4. How many grams are there in:
   - **a** 3 kg?
   - **b** $4\frac{1}{2}$ kg?
   - **c** $12\frac{1}{4}$ kg?
   - **d** 1.5 kg?
5. Copy and complete.
   - **a** 5 cm = ____ mm
   - **b** 3 m = ____ cm
   - **c** 2 km = ____ m
   - **d** 4.5 m = ____ cm
   - **e** 125 mm = ____ cm
   - **f** 4500 mm = ____ m
6. On an analog clock, which number would the long hand point to at the following times:
   - **a** 3.30 p.m.?
   - **b** noon?
   - **c** a quarter to five?
   - **d** ten past 8?
7. Write the following times using 24-hour notation.
   - **a** half past seven in the evening
   - **b** midnight
   - **c** 25 to one in the morning
   - **d** quarter to ten at night

Challenge and investigate
1. Explain what these prefixes mean, and give examples of how we use them in measuring: kilo-, centi-, milli-.
2. **a** Name the unit of length that is one tenth of a centimetre.
   b Explain why we need this unit. Give two examples of objects you would measure in this unit.
3. Explain what we mean by analog and digital clocks. Give examples to explain your answer.
4. **a** A music lesson starts at 2 o'clock and ends at twenty minutes to 3. How long is the lesson?
   b A match started at 4.15 p.m. and lasted for 2 and a half hours. At what time did it finish?
5. The International Space Station takes 644 minutes to orbit Earth 7 times.
   a About how many minutes does it take to complete one orbit?
   b Now write your answer in hours and minutes.

SECTION 5

Chapter 9 Handling data

In this chapter, you will:
- ask questions that can be answered with data
- collect data in different ways
- organise data using tally charts and tables
- draw and interpret different types of graphs
- interpret and analyse data
- find the mode of a data set
- solve problems and make decisions based on data.

Starting point

1 Lana makes jewellery from crystals. She keeps a **tally** of how many round, square and pear-shaped crystals she uses.

Round	ЖΙΙΙΙ
Square	ЖΙΙΙΙ
Pear	Ж Ж ΙΙ

Key word
tally

Answer these questions in pairs.
  a What do the tallies tell you?
  b Which crystal shape has she used most?
  c How many pear-shaped crystals has she used?
  d How can tallies help you organise information?

2 Look at this graph carefully.

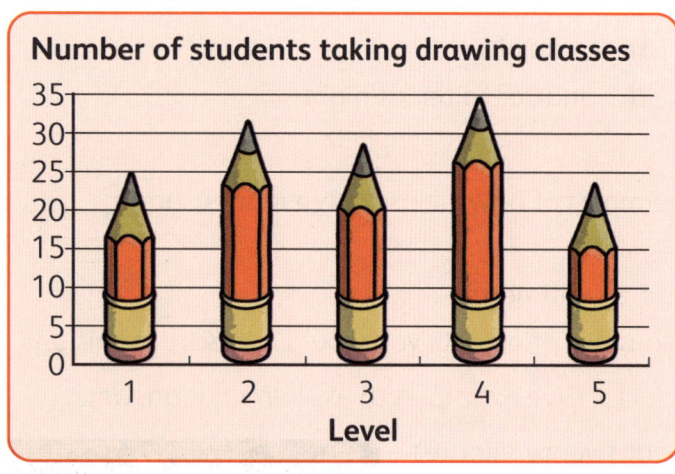

  a Explain why this is not a pictograph even though it has some pictures on it.
  b What type of graph do you think it is? Why?

3 Nigel wonders how the students in his class get to school in the morning.
  a Write the question that Nigel needs to answer.
  b Suggest how he could collect the data he needs to answer his question.
  c How could he organise the data that he collects to make it easier to work with and understand?

Asking statistical questions

Key maths idea

Data is information that you can collect, record and organise to answer a statistical question.

Statistical questions are not simple questions with just one answer. Statistical questions have many different possible answers.

A question such as 'How tall am I?' is not a statistical question because it has only one answer and it is only about you. If you ask 'How tall are the students in Standard 4?', you are asking a statistical question. Your question is about a group and there are different answers because the students are different heights.

Other examples of statistical questions are:
- How much time do children in my school spend online each day?
- What is the most common number of books children my age read per week?

1. Decide whether or not each question is a statistical question. Write Yes or No.
   a. What is the most common colour of car in the mall car park?
   b. Where in town is the biggest mall?
   c. Which subject is the most popular among students in my class?
   d. What type of car does the principal drive?
   e. In which month do we receive the most rainfall?
   f. How long does it take students in our class to get ready for school?
   g. How many books are there in your classroom?

2. Terique asked students in his class what standard they are in. Explain why this is not a statistical question.

Talking maths

Is the question 'Which student in our class travels the furthest to get to school?' a statistical question or not? What do you think? Share your ideas with your partner, giving reasons for your answer.

Collecting and organising data

Key maths idea

There are many ways to collect data.

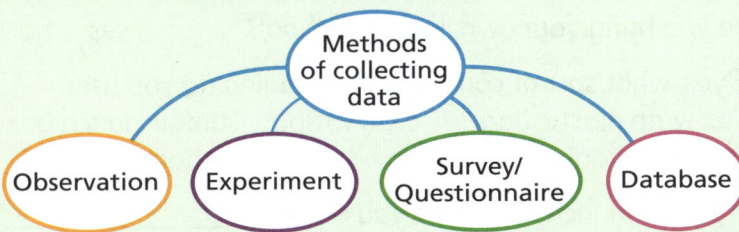

The method you choose depends on the statistical question you are trying to answer.

Section 5 Statistics Chapter 9 Handling data

(continued)

Read each question and how data was collected to answer it.

Question	How data was collected
What colour car is most common in our community?	**Observed** 100 cars passing the school and recorded the colours.
How often do you get a six if you roll a dice 50 times?	Did an **experiment** by rolling a dice 50 times and recording the results.
How can we improve the library facilities at school?	Carried out a **survey** using a **questionnaire** to find out who uses the library and what changes they would like to see.
How has the population of Trinidad and Tobago changed over time?	Used the **database** on the Central Statistics Office (CSO) website to find the data.

Key words
observe / observation
experiment
survey
questionnaire
database

1. Which method do you think is most suitable for collecting the following data? Explain why.
   a. The number of households in Trinidad and Tobago that have internet access
   b. The most popular brand of mobile phone among secondary school students
   c. How much people are prepared to pay for school uniforms
   d. The number of children in your school who wear spectacles
   e. What type of plastic breaks down fastest in sunlight
   f. How people feel about a new road being built in their area
   g. Which type of material will keep water cool the longest

2. List four questions that you could answer by carrying out an observation.

3. Malaika is collecting data. This is her data-collection sheet.
   a. What method has Malaika chosen to collect the data?
   b. What is she collecting data about?
   c. What do you call this type of form?
   d. How should people answer the questions?
   e. What would your answers be?

   School lunch survey
   Please circle the correct information.
   Age: 4   5   6   7   8   9   10
   1. Do you bring your own lunch to school?   yes   no
   2. If yes, what sort of container / packaging do you use?
      wax wrap   plastic bag   cling wrap   tinfoil   lunch box
      ice-cream container   tin   other: _____
   3. If you buy lunch, where do you buy it? _____

Frequency tables

> **Real-life maths**
>
> Data can help people make better decisions. For example, census data helps the government plan and decide where to spend money. Shops collect data about what items sell best so they can plan and make sure they have enough stock. Knowing how to make sense of data is key to being an educated citizen in the modern world.

Tally chart

> **Mental maths**
>
> 1. You can use tallies to keep count without writing numbers or words. Onika and Naomi used **tally charts** to keep track of how many people they served at their coconut-water stalls.
>
> | Onika's stall | Naomi's stall |
> |---|---|
> | ⦀⦀ ⦀⦀ ⦀⦀ ⦀⦀ ⦀⦀ ⦀⦀
 ⦀⦀ ⦀⦀ ⦀⦀ ⦀⦀ ⦀⦀ ⦀⦀
 ⦀⦀ ⦀⦀ ⦀⦀ ⦀⦀ ⦀⦀ ⦀⦀
 ⦀⦀ ⦀⦀ ⦀⦀ III | ⦀⦀ ⦀⦀ ⦀⦀ ⦀⦀ ⦀⦀ ⦀⦀
 ⦀⦀ ⦀⦀ ⦀⦀ ⦀⦀ ⦀⦀ ⦀⦀
 ⦀⦀ ⦀⦀ ⦀⦀ ⦀⦀ ⦀⦀ ⦀⦀
 ⦀⦀ ⦀⦀ ⦀⦀ IIII |
>
> **a** Count in fives to get a total for each person.
> **b** Onika says it is faster for her to count in tens. How can she do that using the tallies?
> **c** How many more people will Naomi need to serve to reach 200?

> **Hint**
>
> Remember: you can use computer applications to organise your data and draw graphs.

Frequency tables

Key words
tally chart
frequency
frequency table

> **Key maths idea**
>
> Once you have collected data, you need to organise it. Tables are very useful for organising data.
>
> This is a tally chart.
>
> | Favourite fast food | Tally |
> |---|---|
> | Roti | ⦀⦀ ⦀⦀ IIII |
> | Doubles | ⦀⦀ ⦀⦀ ⦀⦀ I |
> | Bake and shark | ⦀⦀ ⦀⦀ III |
> | Aloo pie | ⦀⦀ ⦀⦀ ⦀⦀ |
>
> This is a frequency table.
>
> | Favourite fast food | Frequency |
> |---|---|
> | Roti | 14 |
> | Doubles | 16 |
> | Bake and shark | 13 |
> | Aloo pie | 15 |
>
>
>
> **Frequency** is how often something happens. In this example, the frequency is how many times someone chooses each food. This **frequency table** shows the same data as the tally chart.
>
> This table combines tallies and frequencies.
>
> | Favourite fast food | Tally | Frequency |
> |---|---|---|
> | Roti | ⦀⦀ ⦀⦀ IIII | 14 |
> | Doubles | ⦀⦀ ⦀⦀ ⦀⦀ I | 16 |
> | Bake and shark | ⦀⦀ ⦀⦀ III | 13 |
> | Aloo pie | ⦀⦀ ⦀⦀ ⦀⦀ | 15 |

Statistics

97

Section 5 Statistics Chapter 9 Handling data

1 Read the text.

> Sister islands in the Caribbean Sea
> Nestled in the brilliant blue waters
> Golden beaches hug the coasts
> Lush forests drape the inland hills

Vowel	Tally	Frequency
A		
E		
I		
O		
U		

a Copy the table. Use tallies to count the number of times each vowel appears in the verse.

b Complete the table by filling in the frequencies.

2 Patricia has a souvenir stall near the cruise-ship terminal. She wants to know what age groups visit her stall most often. She does a survey of her customers one morning while a ship is docked.

Age group	Tally	Frequency
0–10		3
11–20	ЖΙΙ	
21–30		12
31–40		32
41–50	Ж Ж Ж ΙΙΙ	
51–60	Ж Ж Ж	
> 60	ΙΙΙ	

a Copy the table and fill in the missing tallies and frequencies.

b Work out how many customers she surveyed.

c In which age group were most customers?

d How might this information be useful to Patricia? Share your ideas with your partner.

Problem solving

1 The number of people affected by flight delays of more than 30 minutes on flights to and from Tobago was recorded each day for 40 days. This is the data they collected.

```
23   58   4    9    10   21   33   12   31   11
8    9    56   32   9    11   12   7    11   12
12   24   18   45   51   38   12   16   10   6
9    0    2    1    12   19   102  34   21   7
```

a What makes this data difficult to work with?

b One day, there was a storm and afternoon flights could not take off. Which day do you think this was? Why?

c Copy and complete this frequency table to organise the data.

Number of people	0–10	11–20	21–30	31–40	41–50	51 or more
Frequency						

d Why do you think the number of people has been grouped in tens for this data set?

e How might this data be useful to:
- the airport management?
- passengers?

The mode

Key maths idea

The **mode** is the data value that occurs most often in a set of data.
Priya rolled a die twelve times and got these scores.

We can use a table to show how many times she got each score.

Score						
Number of throws	2	5	1	1	2	1

You can see that she got a score of 2 five times. This is the score she got most often. The mode of her scores is 2.

Key word: mode

1 These are the scores that Sadia and Chin got when they rolled a die twelve times.

Sadia

Chin

What is the mode for each set of scores?

2 a Roll a die 12 times. Find an efficient way to record each score.
b Find the mode of your scores.

Displaying data

Key maths idea

You already know how to draw different types of graphs to display data. The type of graph that you draw depends on your data and who the graph is for.

- **Block graphs** are very simple graphs that are useful for young children. You can draw a block graph if you have a very small set of data.
- **Pictographs** are useful for attracting attention, especially if you choose pictures or symbols that are linked to the data.
- **Bar graphs** are useful for showing the differences between groups in your data. They are a good choice for any audience because they are simple and easy to interpret.

Key words
block graph
pictograph
bar graph

Section 5 Statistics Chapter 9 Handling data

Complete these questions either in pairs or individually.

1. Compare the three graphs. Answer these questions.
   a. What data does each graph show? How do you know?
   b. How is the data shown on each graph?
   c. How do you know what the symbols on the pictograph represent?
   d. Which graph has a scale that shows the frequency?
   e. What is similar about the graphs? What is different?

2. How would you describe each type of graph to someone? Write short notes about the features of each type of graph.

3. An Infant class teacher wants to make a class graph to show how many children are absent each day of the week. What type of graph would be best for this? Why?

Students absent

Mon Tues Wed Thurs Fri

Block graph

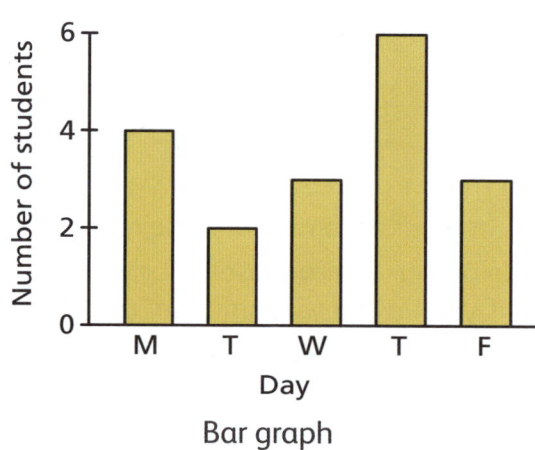

Pictograph

Number of absentees

Bar graph

More about graphs

Key maths idea

Key words
vertical scale
horizontal scale

Bar graphs can be useful when you want to compare different groups or categories in data. For example, this graph shows the number of students (the frequency) that got each grade (the category) in a maths test.

The **vertical scale** shows the frequency. This scale is numbered in tens: 0, 10, 20, 30. Each ten is divided into five smaller intervals of two. So, for example, you can see that only four students achieved a D grade.

The **horizontal scale** is labelled to show the grades. Each bar is the same width and there is an equal space between the bars.

The bar for B is the longest. This tells you that more students got a B than any other grade. B is the mode for this data set.

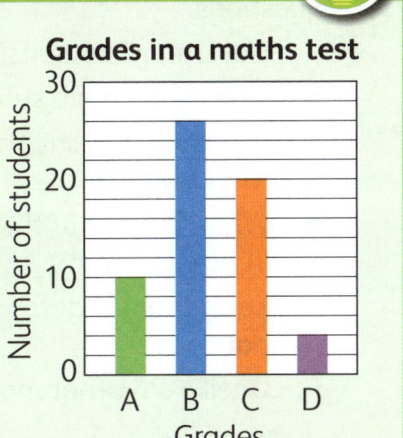

1. When you draw a pictograph, you must provide a key to show what each symbol represents. Josiah used the symbol ⊠ to represent 4 students. How could he show:
   a 2 students?   b 3 students?   c 1 student?

2. Tricia used the symbol ● to represent 20 people. Draw the symbol she could use to represent:
   a 10 people   b 15 people   c 5 people.

3. Karimah did a survey to find how students in her Standard learn about the news. This is her data.
   - 14 students said they watch the news on TV
   - 12 students said they find out about news on social media
   - 10 students said they hear about the news on the radio
   - 6 students said they hear about the news from adults in the community

   a Draw a frequency table to summarise the data.
   b Draw a pictograph to show the data. Choose a suitable symbol and include a key.
   c Draw a bar graph to show the data. Use a scale interval of 4 on the vertical axis.

4. List the first names of 10 students in your class.
   a Make a tally chart to show how many letters there are in each name.
   b Draw a block graph to show your data.

Extension

5. What is the most common shoe size for Standard 4 students?
   a Estimate the answer. Then work in pairs to collect data to find the answer.
   b Organise the data and find the mode. How close was your estimate?
   c How would knowing the most common shoe sizes in each standard help the school supply-store owner to plan?

Talking maths

Have you ever drawn graphs using a computer app? If you have, tell your partner how you did this. If not, find out how you can do this.

Section 5 **Statistics** Chapter 9 Handling data

> **Problem solving** ❓
>
> 1. Use the graph to find the information you need to answer these questions.
>   a. How many students' results are shown?
>   b. How many more students scored B than C?
>   c. Is it correct to say that $\frac{1}{3}$ of the students scored C? Explain why or why not.
>   d. What fraction of the students scored a D? Give your answer in simplest form.
>   e. Keisha says that most students got a B. Explain why this is **not** correct.
>   f. Oneika says that most students scored B or higher. Explain why this **is** correct.

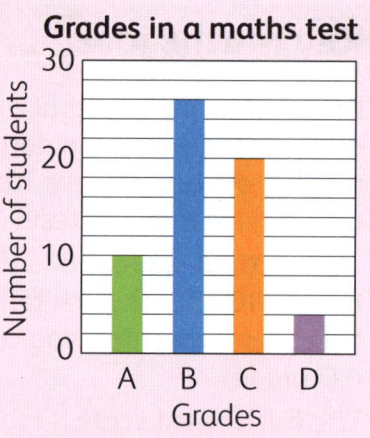

Interpreting graphs

> ### Key maths idea
>
> We read and **interpret** graphs carefully to make sense of the data.
> - Read the headings to find out what the graph is about.
> - Read the labels on the axes to see what data is shown.
> - Make sure you understand what scale is being used on the graph.
> - Look for **patterns** and **trends** in the data.
> - Think about what might be causing the patterns or trends and draw **conclusions** or make decisions based on your interpretation.
> - Sometimes you may be asked to write a paragraph summarising what you can learn from the graph.
>
> **Key words**
> interpret
> pattern
> trend
> conclusion
>
> Look at this example to see how Natasha interpreted and made sense of this data.
>
>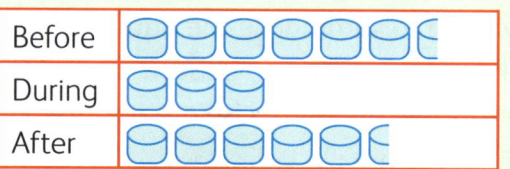
>
> First Natasha made some notes of her own on the graph.
>
>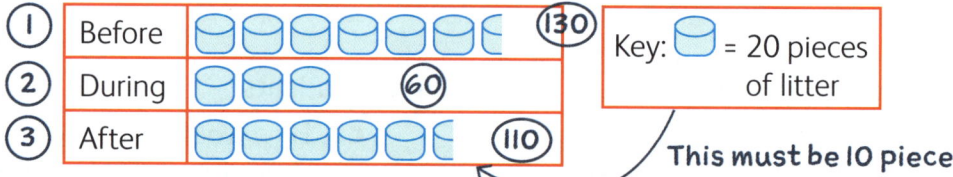
>
> Then she summarised what she learned.
>
> 130 pieces of litter were counted before the campaign. During the campaign, this number went down to 60. This suggests that the campaign worked as people stopped littering.
>
> After the campaign, the number of pieces of litter increased again. I think this means that some people learned from the campaign and stopped littering, but many forgot once it was over. I would suggest that the school have anti-litter campaigns regularly to see whether it makes a bigger difference over time.

Interpreting graphs

1 This graph shows how long students took to wash their hands.

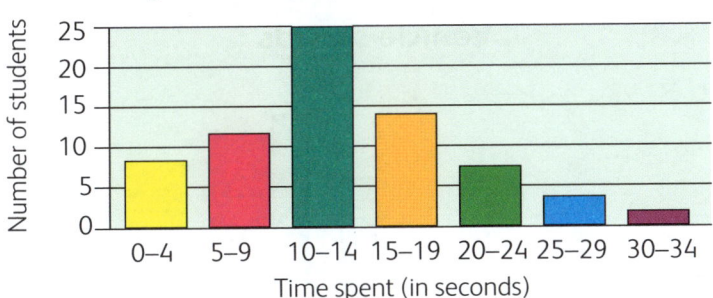

a What is the most common length of time students took to wash their hands?

b How many students spent less than 5 seconds washing their hands?

c The World Health Organization (WHO) guidelines say you should wash your hands for at least 20 seconds. How well does this group of students meet those guidelines?

d Does the graph give you information about students who did not wash their hands? Explain your answer.

e How could you find out whether students at your school meet the WHO guidelines?

2 A group of friends collected data about how many books they read during the summer holidays. Candy and Asha drew bar graphs to display the data.

Candy's graph

Asha's graph

a Asha made one mistake when she drew the bars. What was it?

b Why do the graphs look different even though they show the same data?

c How many books did Candy read?

d Who read fewer books: Asha or Laura?

e How many more books did Candy read than Asha?

f Use the data on Candy's graph to calculate the total number of books read by the five students.

g Suzie says that the bars for her and Laura are the same length so that means they must have read the same books. Explain why Suzie's reasoning is incorrect.

h Nadia says that she actually read the most because she read longer books. How could you check whether Nadia was correct or not?

Section 5 Statistics Chapter 9 Handling data

Problem solving

1. The graph shows the speeds of vehicles travelling along a section of main road.

   a. Which speed was the mode?

   b. Why are there big gaps between the bars?

   c. The speed limit on this stretch of road is 65. How many vehicles were travelling above the speed limit?

   d. A community member says that speeding is not a major problem along this stretch of road. Use data from the graph. Write two statements to support this.

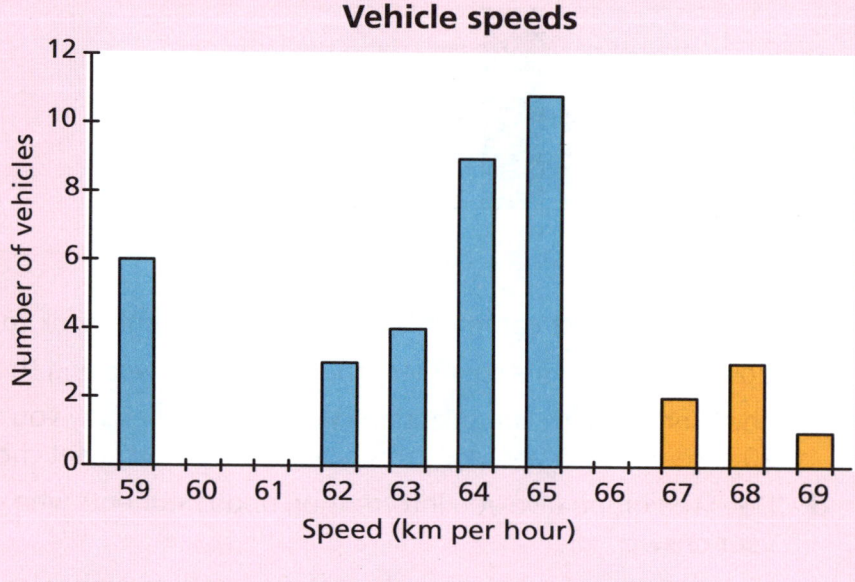

Full STEAM ahead   Make sense of weather data

1. Raj kept a record of the weather every day for a full month. He recorded his data on a calendar sheet.

1	2	3	4	5	6	7
Rainy	Cloudy	Sunny	Sunny	Sunny	Cloudy	Rainy
8	9	10	11	12	13	14
Sunny	Sunny	Sunny	Windy	Windy	Cloudy	Cloudy
15	16	17	18	19	20	21
Sunny	Sunny	Rainy	Rainy	Cloudy	Windy	Sunny
22	23	24	25	26	27	28
Sunny	Sunny	Windy	Cloudy	Rainy	Sunny	Sunny

a. What month was this? How do you know?

b. Draw up a suitable table to organise and display the data more clearly.

c. Draw a pictograph to show how many sunny days there were each week.

d. Draw a suitable graph to compare the number of rainy, windy, cloudy and sunny days for the month.

Working through the data cycle

Key maths idea

Data is important in our lives because it allows us to solve problems and make well-informed decisions. The diagram shows the stages in the data problem-solving cycle.

Most data cycles start with a question or problem. Once you work through the different stages, you may find that you have new questions or problems and so you repeat the stages in an ongoing cycle.

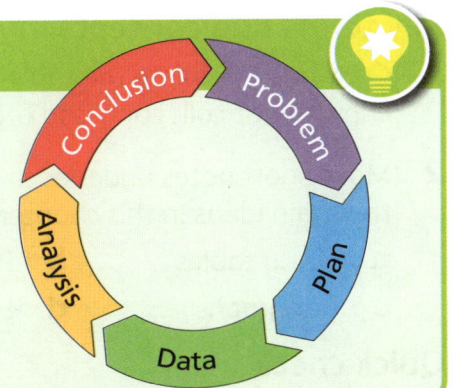

Talking maths

Work in pairs. Talk about what you do in each stage of the process. For example: In the problem stage, we identify a statistical question that needs to be answered.

You are going to work through the data problem-solving cycle by doing your own investigation.

1. Choose a topic that interests you to research at home or at school. You can choose from the ideas below or you can use an idea of your own.

   - climate and weather
   - roads and transport
   - health and exercise
   - environmental issues
   - mobile phones and other devices
   - tourism

2. Follow these steps to complete your investigation.
   a. Write a statistical question for your investigation.
   b. Decide how you will collect the data you need to answer the question.
   c. Collect the data and use a suitable table to summarise and organise it.
   d. Draw two different graphs to display the data. One should be a pictograph and the other should be a bar graph. Use a different scale for each one.
   e. Write a report to explain what you learned from your investigation.
   f. Present your findings to the rest of the class. Decide which data representation will be best to show your findings to your audience.

What did you learn?

Look back at the work you did in this chapter. Rate your progress.
1 = I cannot do this. **2** = I need more practice. **3** = I understand it and feel confident.

Can you:
- ask questions that can be answered by data?
- collect and organise data?
- represent data in tables, pictographs, block graphs and bar graphs?
- interpret data shown in tables, charts and graphs?
- find the mode of a data set?
- draw conclusions and make decisions based on data?

Section 5 Statistics Chapter 9 Handling data

Review: Handling data

Key terms and concepts

1. Copy and complete the table to summarise what you know about collecting and organising data.

Method of collecting data	What you do if you use this method
Observation	
Experiment	
Survey	
Using a database	

2. Make short notes under each heading to summarise the main ideas in this chapter.
   a Using tables
   b Pictographs
   c Bar graphs
   d How to interpret a graph

Quick check

1. A friend asks you 'What is the difference between a statistical question and an ordinary question?' What would you tell them?

2. People collect and use data for different reasons. Give an example of data that each of these people might need and suggest how they could collect and organise the data.
   a A cricket coach in charge of choosing a team
   b A vendor who sells snacks at a primary school
   c A person who wants to buy a new mobile phone
   d A tourist planning a trip to Tobago

3. Explain how two bar graphs showing the same data can look very different from each other.

Challenge and investigate

1. Javid asked people which food they preferred. He gave them these choices:

   (fruit) (vegetables) (rice) (potatoes)
   (fish) (chicken) (roti) (doubles)

   He used this tally chart to record the results.

Food	Tally
Fruit	IIII
Vegetables	IIII I
Rice	IIII II
Potatoes	III
Fish	IIII III
Chicken	II
Roti	IIII I
Doubles	IIII

   a How many people did he survey?
   b Which foods were chosen by an equal number of people?
   c Is it correct to say that more people chose vegetables than fruit?

2. Use the data from Javid's survey to draw a pictograph and a bar chart in your exercise book. Let each symbol on the pictograph represent 2 people and use a scale of 1 cm per two people in the bar graph.

SECTION 6

Chapter 10 Perimeter and area

In this chapter, you will:
- measure and record the perimeter and area of regular and irregular plane shapes
- draw shapes given their area
- work with square metres and centimetres
- solve problems involving area.

Starting point

1. Look at the picture of the football field. Discuss with a partner:
   a. What shapes can you see on the field?
   b. The whole area is covered with grass. The white lines are painted. How do you know how much grass to plant?
   c. How do you know how long to make the lines?
   d. The long side of the pitch is 105 m and the width is 68 m. If you start at one corner and walk all the way around the football pitch, what distance will you walk by the time you get back to the same corner? Explain how you worked it out.
   e. Work out the length and width of each half of the pitch.
   f. How could you find out the distance of the circle at the middle using a piece of string and a tape measure?

Full STEAM ahead Explore area and perimeter

1. Cut a square of paper that is 10 cm long and 10 cm wide. How many of these squares would you need to cover the surface of:
   a. your desk?
   b. the whiteboard or blackboard in your classroom?
   c. a window in your classroom?

   The squares should not overlap or leave any space uncovered.

2. For each of the items above (desk, board, window), guess the distance around it in cm. Then use a ruler or tape measure to measure it accurately. How accurate was your guess?

107

Section 6 Measurement Chapter 10 Perimeter and area

Perimeter

> **Key maths idea**
>
> **Perimeter** is the distance around a closed plane shape. Polygons are plane shapes with straight sides. To find the perimeter of a polygon, you could measure the sides with a ruler or a measuring tape. If you have a sketch with the **dimensions** given, just add them up.
>
> A **regular** shape has all its sides equal in length. So, if you know a shape is regular, you can multiply the length of one side by the number of sides.
>
> For curved sides, you can lay a piece of string along the line and then measure the string.
>
> **Key words**
> perimeter
> dimension
> regular

> **Mental maths**
>
> 1 What is the perimeter of a square if each side is:
>   a 1 cm long?   b 2 cm long?   c 3 cm long?
> 2 What is the rule for calculating the perimeter of a square?
> 3 A rectangle always has both pairs of opposite sides equal in length. Work out the perimeter of a rectangle with:
>   a width 3 cm and length 4 cm   b length 7 m and width 5 m
> 4 What is the rule for calculating the perimeter of a rectangle?

1 Use a ruler to measure the sides of each shape in centimetres and millimetres. Add them up to work out the perimeters.

   a    b 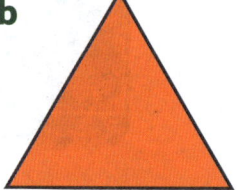   c

2 Calculate the perimeters of the shapes described below. Write a sentence to explain the strategy you used.
   a a square with each side 2 m long
   b an equilateral triangle with each side 4.5 cm long
   c a rectangle with length 2.2 cm and width 1.5 cm

Examples of a strategy: adding up the side lengths, or multiplying the lengths by the number of equal sides.

3 Discuss with a partner how you worked out your answers to question **2**.
4 Use squared paper. Draw the following squares:
   a a square with perimeter 16 cm   b a square with perimeter 5 cm
5 Use squared paper. Draw:
   a two different rectangles that both have a perimeter of 12 cm
   b three different rectangles that each have a perimeter of 24 cm.

Calculating perimeter

Problem solving

1 Ria says that the perimeter of a square is always a square number. What is her mistake?
2 Draw an example of a square whose perimeter is a prime number.

Calculating perimeter

Key maths idea

Sometimes you will be given a sketch that gives the dimensions of a shape. You can use these measurements to work out the perimeter.

Example 1

8 m + 3 m + 7 m = 18 m

The perimeter of the triangle is equal to 18 m.

Example 2

On this shape, the lengths are in different units.

It is easier to find the perimeter when you are working with the same units.

65 cm = 0.65 m

$\frac{1}{2}$ m = 0.5 m

0.8 m + 0.65 m + 0.5 m = 1.95 m

The perimeter of this triangle is equal to 1.95 m.

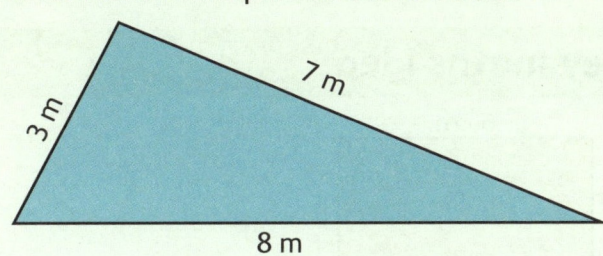
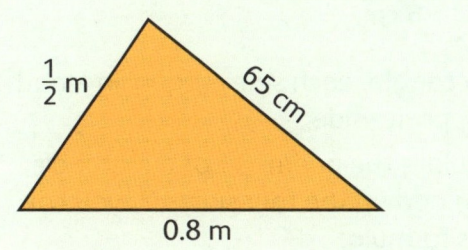

1 Calculate the perimeters of these irregular polygons.

a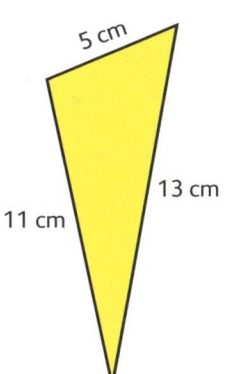

b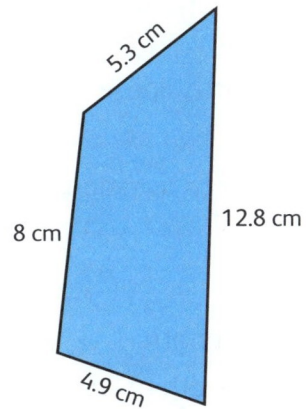

c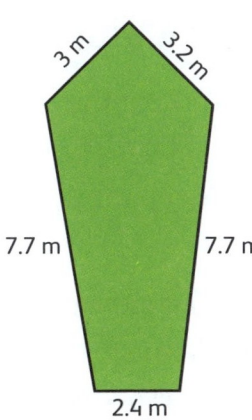

d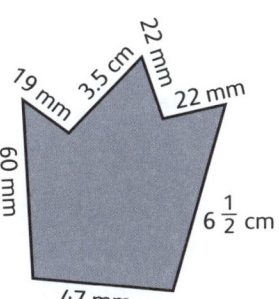

e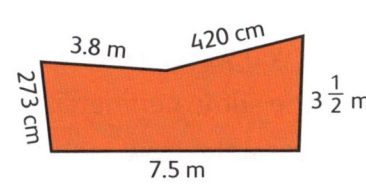

f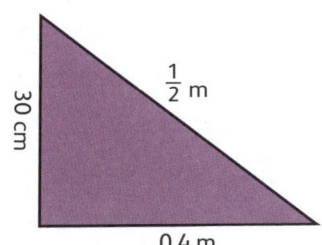

Section 6 Measurement Chapter 10 Perimeter and area

2 Explain what you did to work out the perimeter of shapes **e** and **f** in question **1**.

3 Use the measurements given on these sketches. Calculate the perimeter of each rectangle.

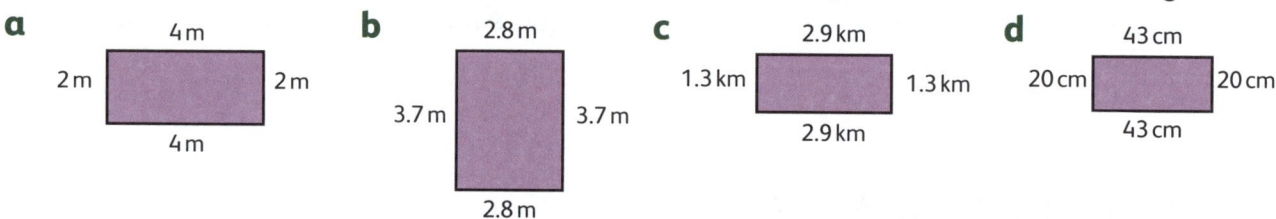

4 Look at your work from question **2**.
  a Can you suggest a shortcut for calculating the perimeter of a rectangle?
  b A square has four equal sides. Write a rule for calculating the perimeter of a square.

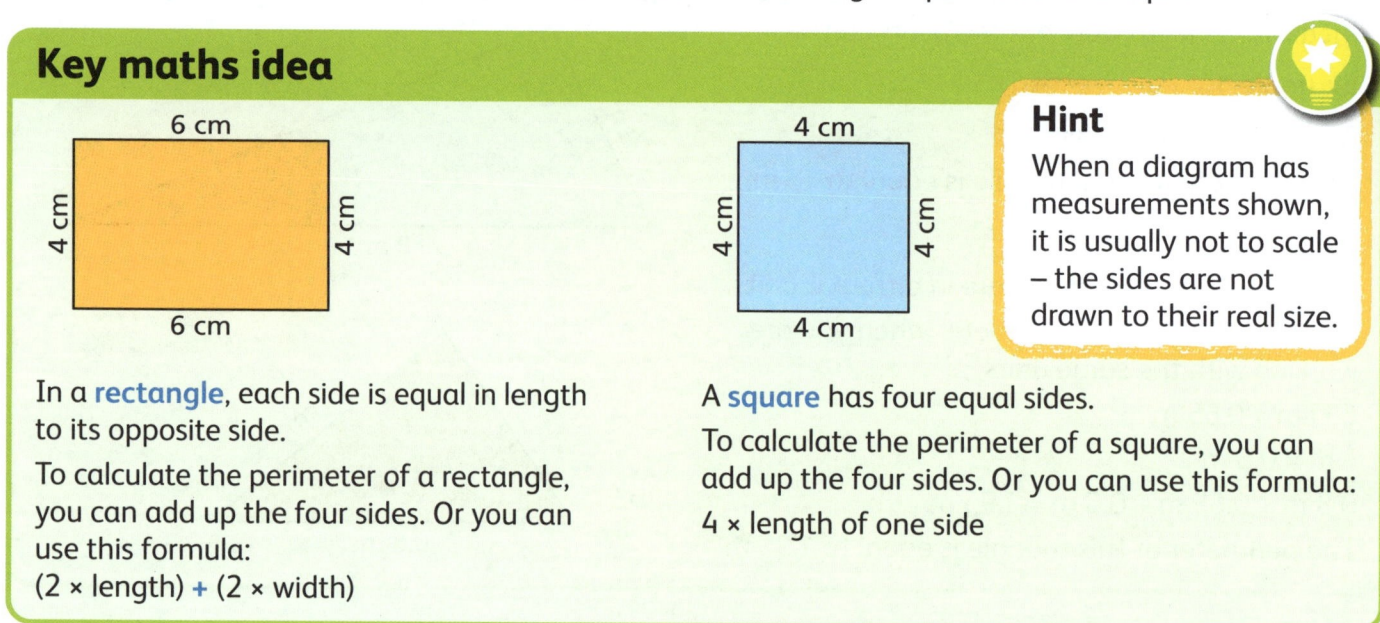

Key maths idea

In a **rectangle**, each side is equal in length to its opposite side.

To calculate the perimeter of a rectangle, you can add up the four sides. Or you can use this formula:
(2 × length) **+** (2 × width)

A **square** has four equal sides.

To calculate the perimeter of a square, you can add up the four sides. Or you can use this formula:
4 × length of one side

Hint

When a diagram has measurements shown, it is usually not to scale – the sides are not drawn to their real size.

Mental maths

1 Work out the perimeter of a square with each side length:
  a 1 cm   **b** 2 cm   **c** 3 cm   **d** 4 cm   **e** 5 cm

2 What pattern do you notice in your answer to question **1**?

1 Calculate the perimeters of each item.
  a A square cake with each side 20 cm in length
  b A pillow with a length of 60 cm and a width of 45 cm
  c A rectangular window with a length 2.4 m and a height 0.5 m

2 a Draw a square with a perimeter of 20 cm.
  b Write a sentence explaining how you worked out the side lengths.
  c Now draw a rectangle that has the same perimeter (20 cm).
  d Compare your answers with a classmate. How many different solutions did you find?
  e Jabari says that you can draw different rectangles with the same perimeter, but squares that have the same perimeter will always be exactly the same size and shape. Explain whether this is true, and give a reason.

Perimeter

Problem solving

1. A poster board has a length of 90 cm and a width of 65 cm. Sadia needs enough tape to make a border all the way along the edge of the poster. Calculate the perimeter of the board to work out how much tape she needs.

2. Ashanti makes a square cake with each side 24 cm in length. What length of ribbon would she need to go around the cake?

3. A photograph is 16 cm wide by 20 cm long. Kimani works out the perimeter, plus an extra 1.5 cm on each end of each side, to work out how much tape he needs to stick the photograph in his album. What is the total length of the tape?

4. The perimeter of a square cushion is 160 cm. What is the length of each side?

5. Look at this diagram of an Olympic-sized swimming pool.

    The pool is 50 m long and 18 m wide.

    a. What is the perimeter of the pool?

    b. Each lane is 2.5 m wide. What is the perimeter of one of the lanes?

    c. There is a non-slip area 50 cm wide on each side of the pool. What is the outside perimeter of the non-slip area?

6. A five-sided flower bed has a perimeter of $17\frac{3}{4}$ metres. There are two sides of $3\frac{1}{4}$ metres long and two sides of $2\frac{1}{2}$ metres long. How long is the other side?

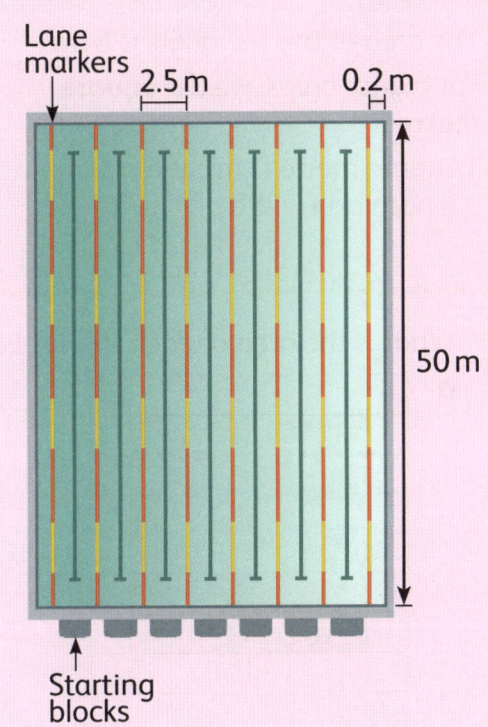

Full STEAM ahead — Measure perimeters at home

1. Find each object below in your classroom or at home. Use a ruler, a tape measure or a metre rule. For each object, measure the height and width in cm, then use it to work out the perimeter.

You will need:
- a ruler
- a tape measure or a metre rule.

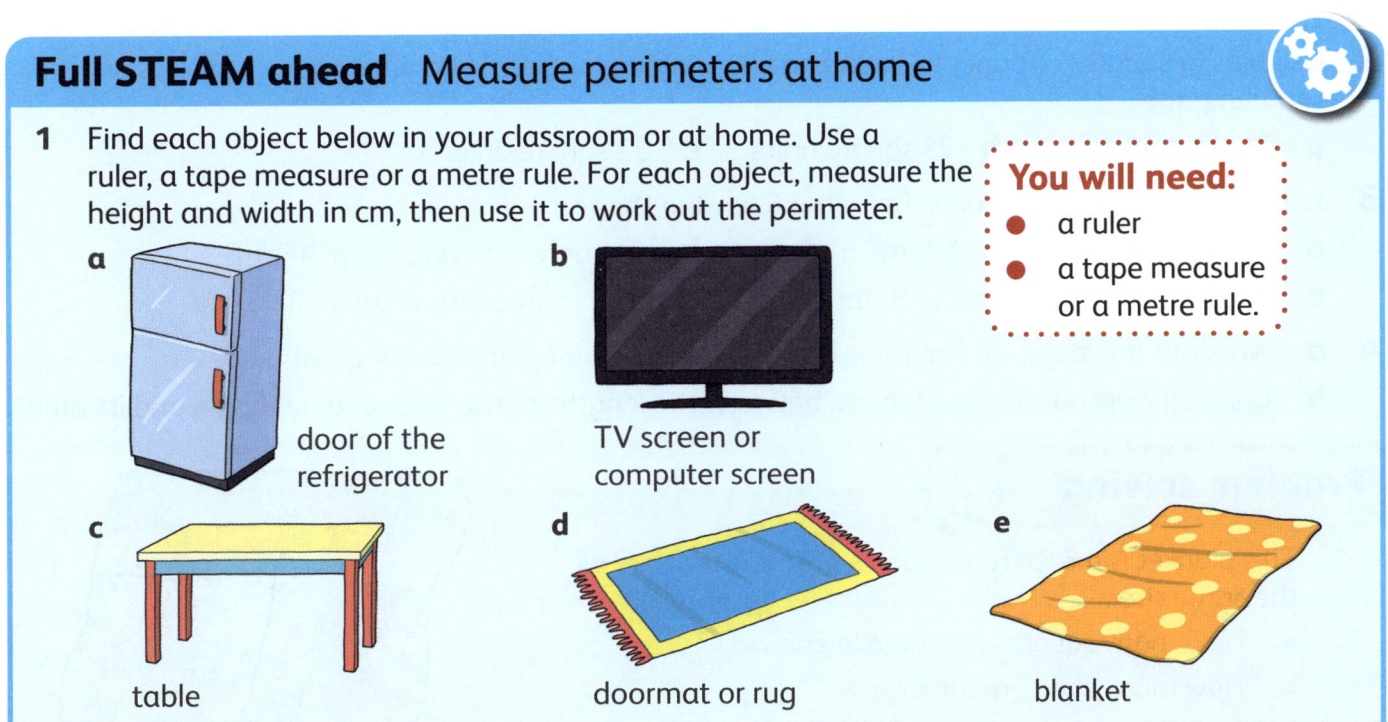

a. door of the refrigerator
b. TV screen or computer screen
c. table
d. doormat or rug
e. blanket

Section 6 Measurement Chapter 10 Perimeter and area

Area

Key maths idea

Area is a measure of how much space a flat shape takes up. You measure area in **square units**.

1 **square centimetre** (1 cm²) is a square with a length and width of 1 cm.

The area of this rectangle is 6 cm².

For bigger areas, we use **square metres**.

1 square metre (1 m²) is a square with a length and width of 1 m.

Key words
area
square unit
square centimetre
square metre

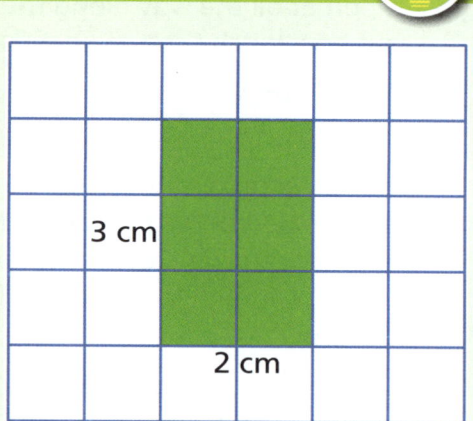

1 What is the area of each shape? Each square represents 1 cm².

   a 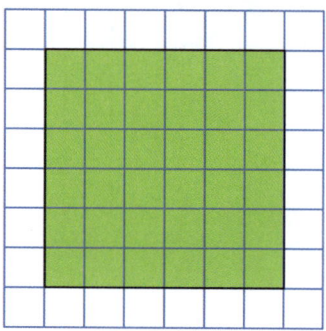   b 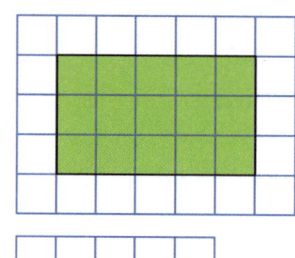   c

   d 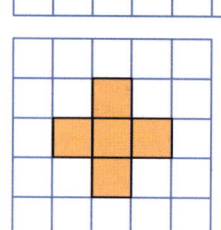   e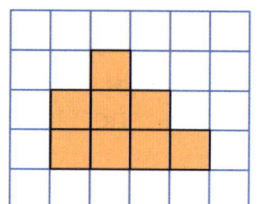

2 Use tiles or squares of paper to make the following rectangles. Then draw your rectangles on squared paper.

   a 24 square units   b 28 square units   c 32 square units

3 Use centimetre-squared paper. Draw the following shapes:

   a a square with an area of 9 cm²   b a square with an area of 16 cm²
   c a rectangle with an area of 8 cm²   d a rectangle with an area of 15 cm²

4 a Work out the length of each side of the rectangles that you drew for question **3**.
   b Can you work out the relationship between the lengths of the sides of a rectangle and its area?

Problem solving

1 This blanket is made from squares of fabric that are all the same size.
   a How many squares are there in each row?
   b How many rows are there?
   c If each square has sides of 10 cm, what is the total area of the blanket?

Areas of irregular shapes

Key maths idea

This is an **irregular shape**. We cannot use a rule to find its area, but we can use a grid of squares to estimate the area of the shape.

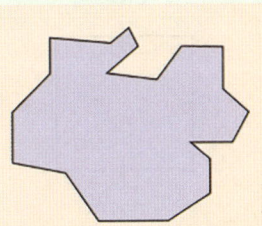

The shape covers some but not all the blocks on the grid.

Each block = 1 cm²

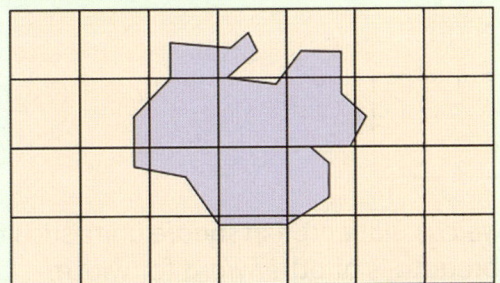

Count all the blocks fully covered by the shape. For all the blocks partly covered, round up to 1 block if there is more than one half covered, or down to 0 if there is less than one half covered. Then work out the total number of blocks covered to give the estimated area.

Key words
irregular shape

1 Estimate the area of each white shape. Each square in the grid represents 1 cm².

   a 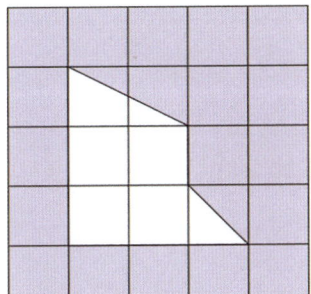   b 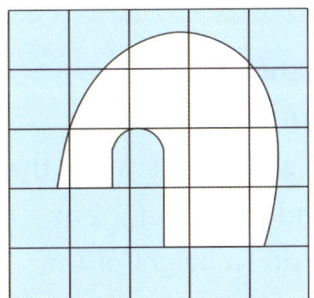   c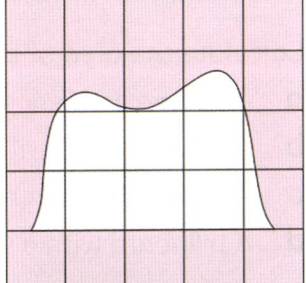

2 Collect some leaves from outside. Estimate the area of each leaf in cm². Then place them on squared paper and trace their outlines. Use what you learned here to work out the area of each leaf.

Problem solving

1 This map shows the areas of some islands. Island A has an area of approximately 7 square units.

   a Which other island has an area of more than 4 square units?

   b Work out the difference in area between the smallest island and the biggest island.

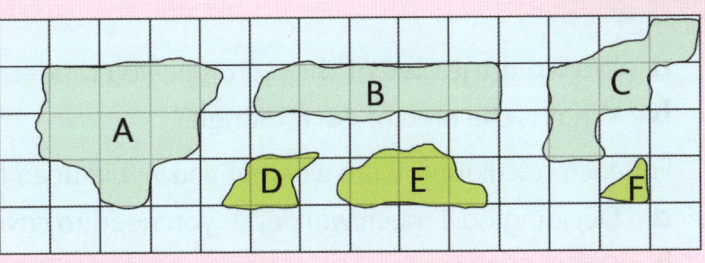

Section 6 Measurement Chapter 10 Perimeter and area

Calculate area

> **Key maths idea**
>
> You can calculate the area of a rectangle using this formula:
>
> area = length × width
>
>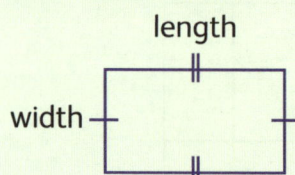
>
> A square has all its sides equal in length, so the formula is:
>
> area = side × side
>
> Or area = (side)^2
>
>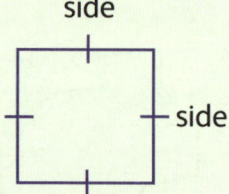
>
> Remember: we calculate area in square units such as square centimetres (cm^2).
> Remember: breadth is another word for width.

> **Mental maths**
>
> 1 Square numbers are the product of a number multiplied by itself, for example 1 × 1 = 1, 2 × 2 = 4, and so on. With a partner, say the first ten square numbers.
>
> 2 For each square number you said, say the side length that would produce that area.

1 Calculate the area of the following shapes.
   a A square with each side equal to 9 m
   b A piece of land with the length equal to 3 km and the width equal to 2 km
   c A wall with a height of 1.8 m and a width of 3.2 m
   d A billboard with a width of 8 m and a height of 4 m

2 Work with a partner. Draw accurate versions of these two triangles. Tape your triangles together to create a rectangle.

   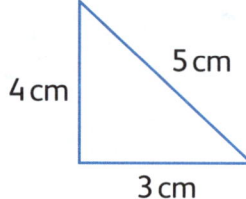 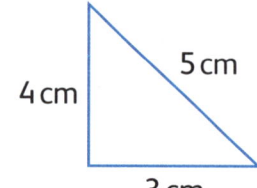

   Hint
   You doubled the triangles to make the rectangles.

   a Work out the area of the rectangle you made.
   b What is the area of each triangle?

3 For each calculation, write whether you would need to find the area or perimeter to solve the problem.
   a Deciding how much wallpaper you need to cover a wall
   b Ordering the correct length of fencing needed to go around a field
   c Calculating how much wood to use for a picture frame
   d Finding out how many patches of roll-on grass are needed to fill a garden bed

4 Use centimetre-squared paper. Make three different shapes that have:
   a an area of 24 cm^2
   b a perimeter of 8 cm.

Areas of compound shapes

5 Write whether each statement is true or false. Use your work from question 4 to help you explain why.
   a Two shapes can have the same area but different perimeters.
   b Two shapes can have the same perimeter but different areas.
   c A shape can have the same perimeter as its area.

Problem solving

1 Look at this scarf and read the measurements.

   A: 200 cm   B: 2100 cm²

   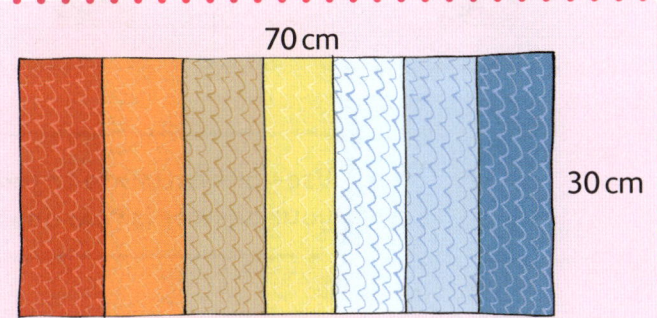

   a Which measurement is the perimeter of the scarf: A or B?
   b Which measurement is the area of the scarf: A or B?
   c What is the difference?
   d How did you calculate each measurement?

2 Deborah is making a blanket for her baby sister's bed. She will make squares and join them together to make the blanket. The finished blanket will be 40 cm wide and 80 cm long.
   a What should each square measure?
   b What is the area of each square?
   c How many squares does Deborah need to make for the whole blanket?

Areas of compound shapes

Key maths idea

A **compound shape** is built up of other shapes. There are three methods you can use to find the area of a compound shape:
- Count the square units.
- Add up the areas of the shapes that make it up.
- Subtract the parts of the shapes that are missing.

For example, for shapes A, B and C, you can simply count up the squares.

For shape D, you need to use what you know about squares and rectangles.

Area of the whole rectangle: 8 m × 12 m = 96 m²
Area of the cut-out part: 2 m × 2 m = 4 m²
Area of the compound shape: 96 m − 4 m = 92 m²

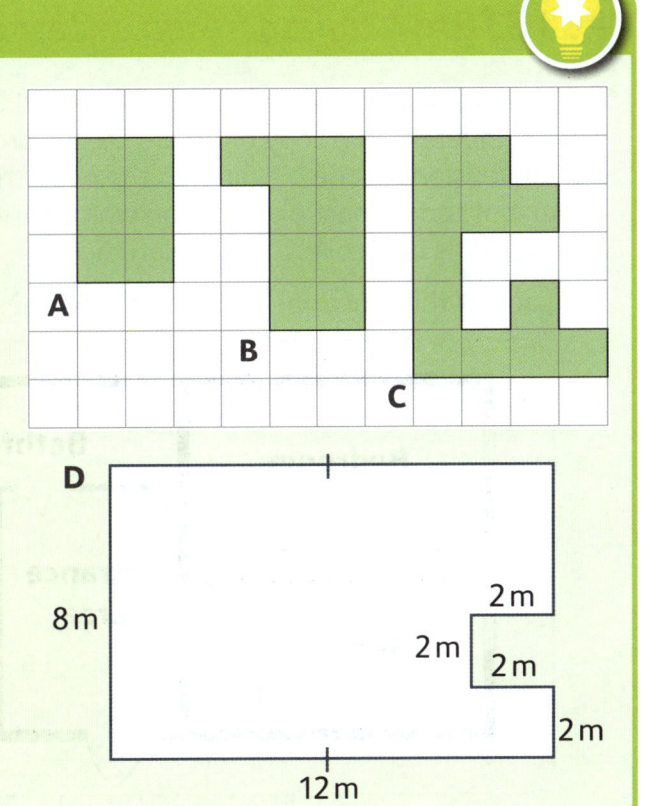

Key words
compound shape

Section 6 Measurement Chapter 10 Perimeter and area

1 On the plans below, each square represents one square centimetre. Estimate the area of each shape. Then calculate the actual area.

a   b   c   d

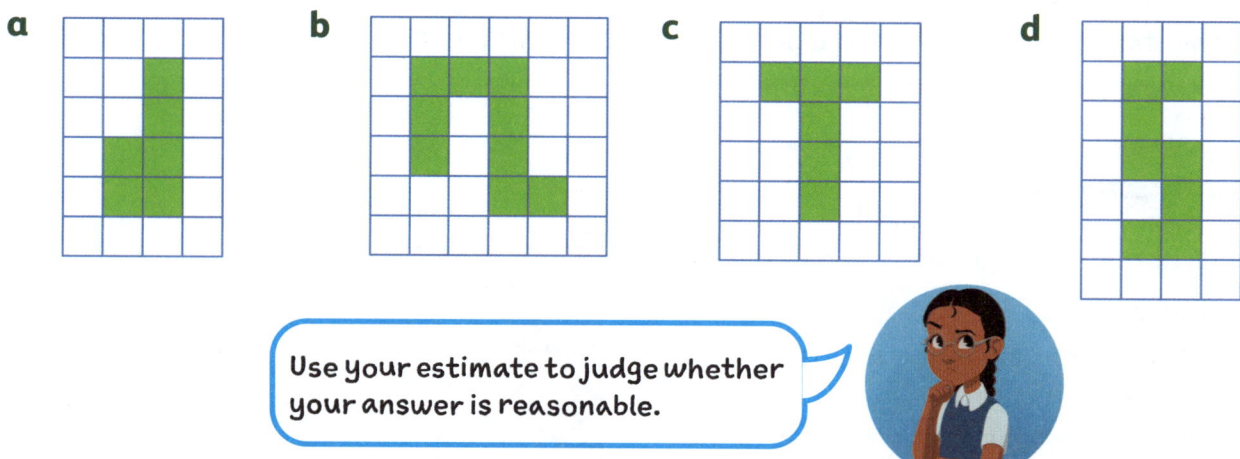

Use your estimate to judge whether your answer is reasonable.

2 Estimate the area of these compound shapes. Then calculate the area of each shape.

a 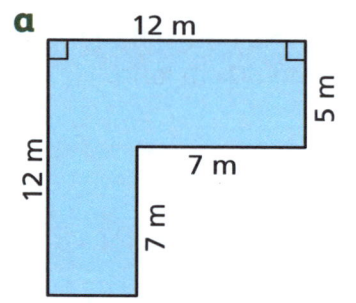   b 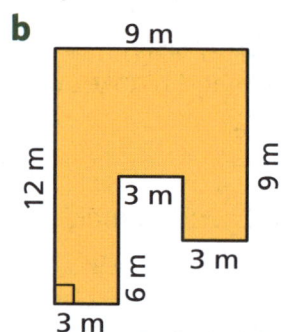   c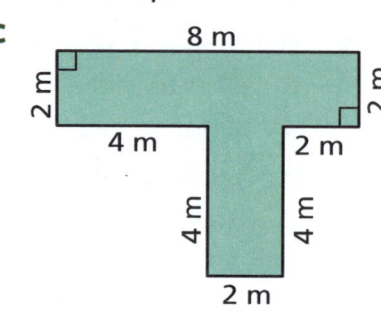

Full STEAM ahead   Area in design

Designers, builders and architects all use perimeter and area in their daily work. A blueprint is a copy of an architect's drawing, showing the details and dimensions of the building. In the past, the copy was printed on blue paper, but today most plans are on white paper. A blueprint is also called a floor plan.

Key words
compound shape

1 Look at this blueprint and answer the questions.

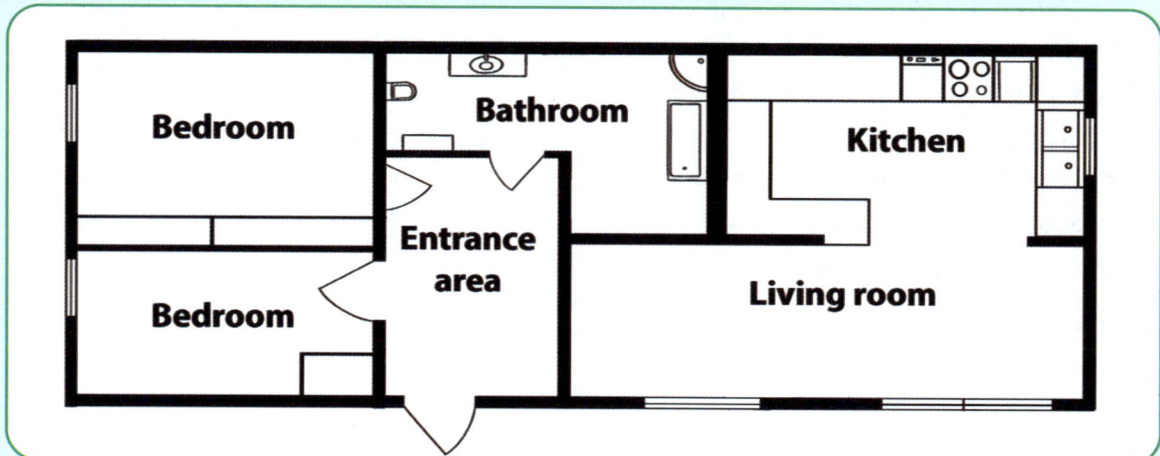

Areas of compound shapes

(continued)

a How many bedrooms are there?
b Which is the biggest room in the house?
c Which two spaces are roughly the same size?
d If the word 'Kitchen' was not on the plan, how would you know that this part of the plan shows the kitchen?
e If the word 'Bathroom' was not on the plan, how would you know that this part of the plan shows the bathroom?

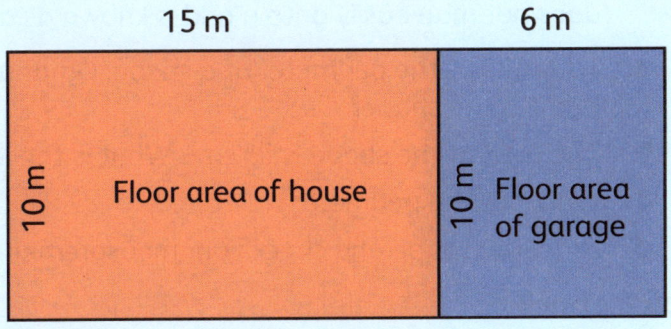

2 The picture shows a house with a garage attached. The plan gives you the dimensions of the house and the garage.

a Calculate the perimeter of the house only.
b Calculate the perimeter of the garage only.
c What is the total perimeter measurement of the house and garage together?

3 a Use grid paper to create a drawing or plan of an imaginary building that you would like to own. Get creative and think of buildings such as a dog's house, a doll's house, a treehouse or a home you might want to live in one day.
   b Once your drawing is complete, count the squares and work out the perimeters and areas of the walls and rooms in your house. Write these in your exercise book.

What did you learn?

Look back at the work you did in this chapter. Rate your progress.
1 = I cannot do this. **2** = I need more practice. **3** = I understand it and feel confident.

Can you:
- measure and record the perimeter and area of regular and irregular plane shapes?
- draw shapes given their area and perimeter?
- work with square metres and centimetres?
- solve problems involving area and perimeter?

Section 6 Measurement Chapter 10 Perimeter and area

Review: Perimeter and area

Key terms and concepts

1. Write the terms missing from each sentence.
   a. ____ is the distance around a closed shape. To calculate it, you add up the lengths of all the ____.
   b. When you work out how many squares of space would cover a flat plane shape, you are measuring ____. We use units such as ____ or ____ to measure this.
   c. A plane shape composed of two or more other shapes is known as a ____ shape. A shape that does not map easily onto a grid is known as an ____ shape.

2. a. To work out the perimeter of a shape, Omar multiplies the length of one side by four. What is the shape?
   b. The area of the shape is 25 cm². What is the length of each side?
   c. Work out the perimeter.
   d. Draw the shape and check your measurements with a ruler.

Quick check

1. For each item, write whether you would need to find the area or the perimeter.
   a. How much carpet you need to cover the floor of a bedroom
   b. The length of fencing you need to enclose a yard
   c. The distance around a swimming pool
   d. The tiles it takes to cover the floor of a room

2. Draw squares with the following side lengths. Work out the perimeter and area of each square.
   a. Sides of 2 cm   b. Sides of 3 cm   c. Sides of 4 cm

3. A rectangle has a perimeter of 20 m. What are the possible lengths of its sides, in whole numbers?

4. For each of the following calculations, write the formula you would use, and give an example:
   a. the perimeter of a square
   b. the perimeter of a rectangle
   c. the area of a square
   d. the area of a rectangle

Challenge and investigate

1. A square has each side equal to 8 cm. Calculate:
   a. the perimeter
   b. the area

2. A rectangular vegetable plot is 8 m long. It has an area of 40 square metres. The farmer increases the length of the plot by 2 m. He wants the new plot to have an area of 60 square metres. By how much does he need to increase the width?

3. Sharon has 18 metres of rope. She wants to mark out a rectangle with the greatest possible area using the rope. The sides of the rectangle must be lengths of whole numbers. Work out the length and width of the rectangle she should make.

SECTION 6

Chapter 11 Capacity and volume

In this chapter, you will:
- understand and work with volume and capacity
- work with non-standard units, such as spoons, cups and boxes
- work with cubic units, litres and millilitres
- show an understanding of the relationship between volume and capacity
- solve problems involving volume and capacity.

Starting point

1. Discuss the picture.
   - Which containers take up the most space?
   - What do you notice about the way that they stack?
   - Which containers leave the least space between them when you pack them on the shelf?

2. Think of containers that your family uses at home.
   - What kinds of items would you store in the bigger containers?
   - What kind of things would you store in the smaller containers?

   Give reasons for your ideas.

Solid objects occupy space

Key maths idea

Solid objects, such as cubes and cuboids, are solids. They have three **dimensions**: length, width and depth. This is why they are called **three-dimensional (3-D)** solids. They occupy space. When we measure how much space a solid takes up, we are measuring its **volume**.

Key words
dimensions
three-dimensional (3-D)
volume

1. Make a collection of boxes of different volumes. Give the boxes letters so you can identify them. Then try to answer these question about your collection:
   a. Write the letters of three boxes with the biggest volumes.
   b. Write the letters of the three boxes with the smallest volumes.
   c. Identify a box that could fit more than four times into one of the other boxes. Explain how you decided.

Section 6 Measurement Chapter 11 Capacity and volume

Using cubic blocks to build solids

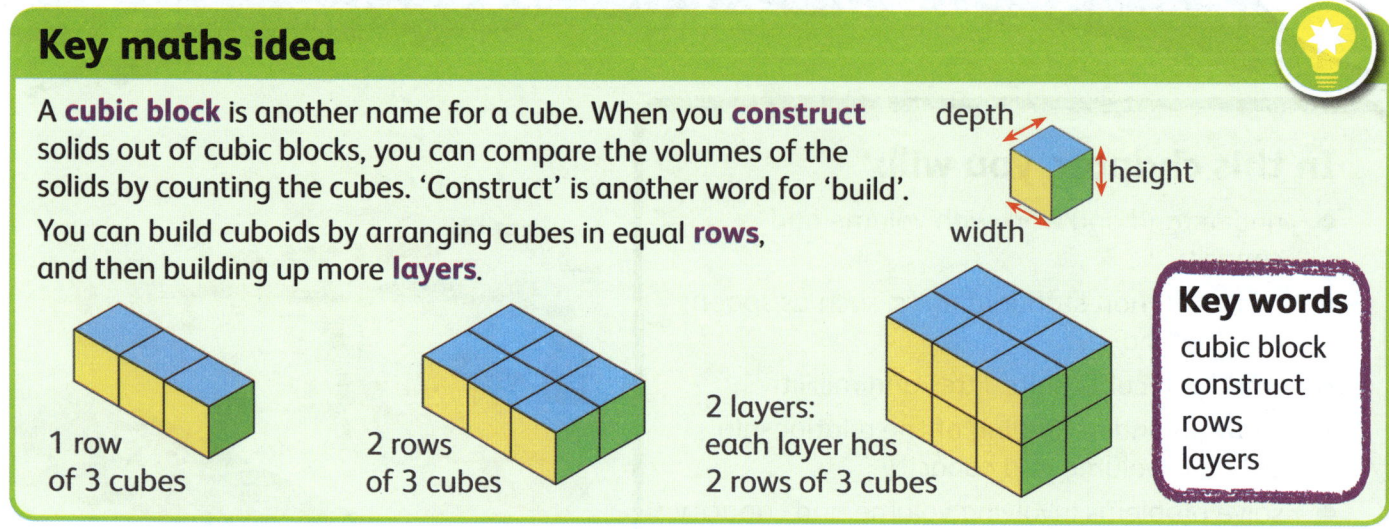

Key maths idea

A **cubic block** is another name for a cube. When you **construct** solids out of cubic blocks, you can compare the volumes of the solids by counting the cubes. 'Construct' is another word for 'build'.

You can build cuboids by arranging cubes in equal **rows**, and then building up more **layers**.

1 row of 3 cubes

2 rows of 3 cubes

2 layers: each layer has 2 rows of 3 cubes

Key words
cubic block
construct
rows
layers

1   How many cubes were used to build each shape? If you have cube blocks in your classroom, check your answer by building each shape.

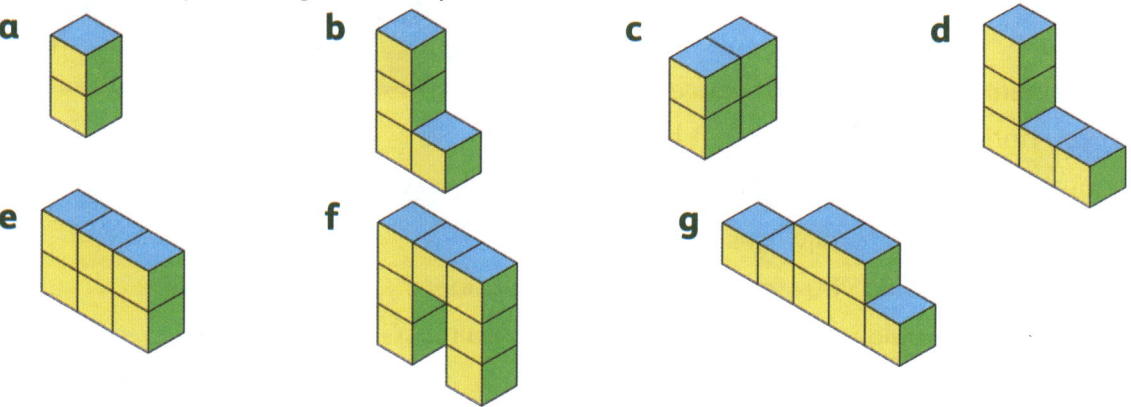

2   True or false? Use the shapes above to explain your answer.
   a   One box can be longer than another, even if they have the same volume.
   b   A shape constructed from four cubes must be taller than a shape constructed from two cubes.

3   How many cubes were used to construct each solid?

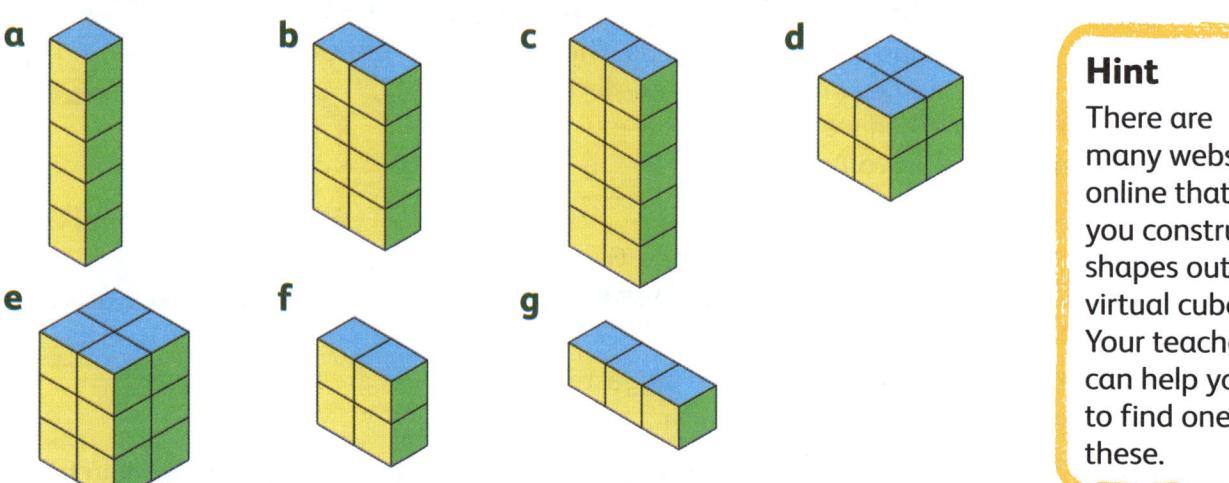

Hint
There are many websites online that let you construct shapes out of virtual cubes. Your teacher can help you to find one of these.

4   Use real or virtual cubes to construct five different cuboids out of cubic blocks. Arrange them in order from the smallest to the greatest volume.

Packing cubes into boxes

Key maths idea

3-D solids have length, width and depth. They occupy space. Even though most objects in real life are not box-shaped, it is helpful to think about 3-D space as blocks or cubes.

Example

The red lines show the shape of a box. How many blocks would you need to fill the box?

You can see there are already some blocks in the box that show us its length (how long it is), width (how wide it is) and height (how high it is).

The box is 4 blocks long and 3 blocks wide.

It takes 12 blocks to fill one layer.

The box is 2 blocks high. So it will have 2 layers when it is full.

12 × 2 = 24

It will take 24 blocks to fill the container.

1 Work out how many blocks you can fit into each box.

a 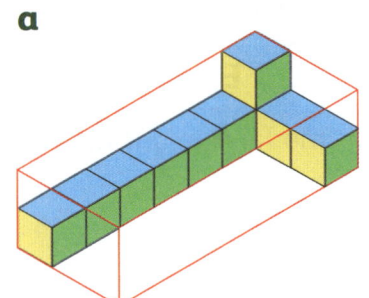　　b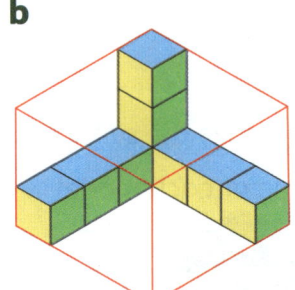

2 How many blocks with the same volume as solid A could you pack into a box with the same volume as solid B?

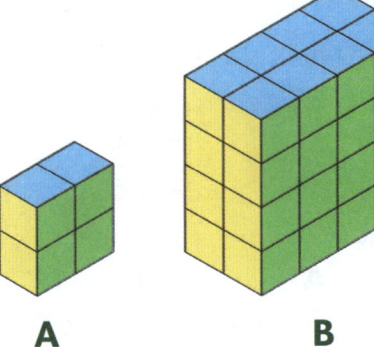

A　　　　B

Hint
Build models with blocks if you need help working it out.

Full STEAM ahead   Packing boxes

Work as a group.

Estimate how many of the smaller boxes you can fit into the larger box, leaving as little space as possible.

One student should pack the boxes into the larger box.

The rest of the group should discuss whether the boxes could be arranged differently.

Take turns trying different arrangements.

How many tries does it take to fit the greatest number of boxes with the least space wasted?

You will need:
- several small cardboard or plastic boxes that are cuboids
- a larger packing box, cake box or plastic storage container.

Section 6 Measurement Chapter 11 Capacity and volume

Volumes of irregular solids

> **Key maths idea**
>
> The volume of a shape is the amount of space it takes up.
>
> This cuboid is made up of 2 rows of 3 blocks. It takes up 6 blocks of space.
>
> This cuboid also takes up 6 blocks of space, even though it is a different shape. It is not a regular cuboid. To work out the volume, count the cubes you can see. Another way to do this is to think of the shape as an incomplete cuboid. Imagine two rows of 5 cubes (2 × 5 = 10), and then take away the 4 that are missing from the second row (10 − 4 = 6).

1 Natasha built some models with wooden blocks.
   a Work out the volume of each model in blocks.
   b Which models have the smallest volume?
   c Which has the greatest volume?
   d What is the difference between the greatest and smallest volume?

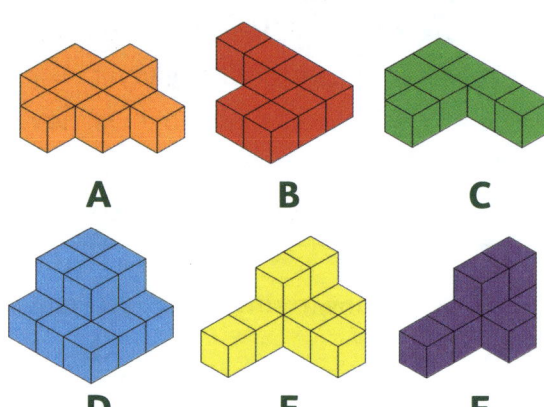

2 Lara says that if objects are different shapes, their volumes must be different. Use Natasha's models to explain why Lara is incorrect.

3 Draw or build two different models that would have the same volume as model **A** in question **1**.

4 Safraz built these models. Work out the volume of each model in cubes.

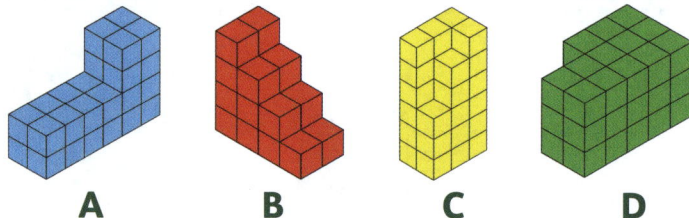

5 Erica has some plastic boxes of different sizes. How many cubes will she need to fill each box?

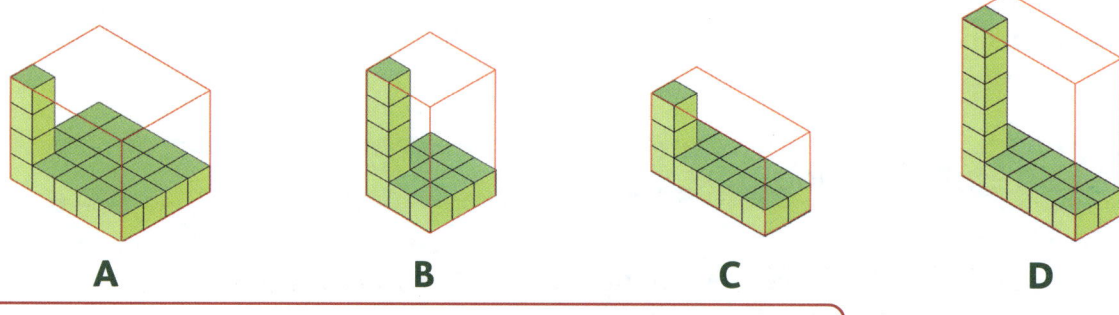

> **Extension**
>
> 6 Reza wants to build this shape. How many cubes does he need?
>
>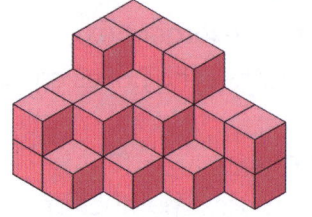

122

Volume in cubic units

Key maths idea

The **volume** of a solid is the amount of space that the solid takes up. We measure volume in **cubic units**. These are cube shapes that are constructed from standard units such as centimetres or metres.

A **cubic centimetre** (1 cm³) has all its edges 1 cm long.

A **cubic metre** (1 m³) is a cube that has all its edges 1 m long.

This cuboid has a volume of 12 cm³. It has 2 layers. Each layer has 6 cubes.

Using standard units of volume means that people understand each other when they give measurements of different spaces. For example, manufacturers need to make sure their products can be packed and transported in boxes and containers that fit.

Key words
volume
cubic unit
cubic centimetre
cubic metre

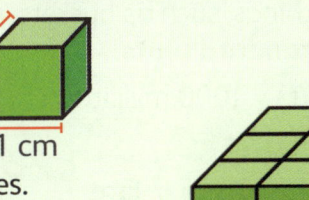

1. Each solid below is built from cubic centimetre blocks. Write the volume of each solid in cm³.

   a    b    c 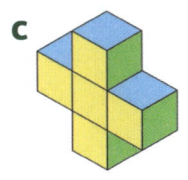   d    e

2. For each of the solids **b** to **e** above:
   a. What is the smallest number of cubes you would need to add, without rearranging the existing cubes, to turn it into a solid cuboid?
   b. Write the volume in cm³ of each new cuboid.

3. Use strips of wood, or strong wire, and tape. Build a model of 1 cubic metre, like the one in the picture to the right.

4. a. How many students do you think can fit into the cubic metre (standing up)? Discuss and estimate as a class.
   b. Let one student at a time carefully step inside the cubic metre model, until no more can fit. The students inside should cover their eyes, and keep guessing how many more can fit.
   c. How well did you estimate?

5. Look again at the solids in question **1**. Imagine that the cubes in the diagrams each represent 1 cubic metre. In groups, discuss whether each solid would fit into your classroom. Use your body to show how much space you think it would occupy in the classroom.

6. Choose one of these real-life situations, or think of a different real-life situation where people use volume measurements. Write why you think it would be important for this person to use standard units of volume.
   - A builder ordering concrete for the floor of a new building
   - A garden designer ordering soil for a hotel garden
   - An engineer designing a storage container for fuel

Section 6 Measurement Chapter 11 Capacity and volume

Litres and millilitres

Key maths idea

The **standard units** for measuring capacity are **litres** and **millilitres**. When we measure using other things, such as buckets, we are using **non-standard units**.

1 litre (ℓ) = 1000 millilitres (ml)
$\frac{1}{2}$ ℓ = 500 ml
1 cup = 250 ml = $\frac{1}{4}$ litre

When you buy drinks such as juice, milk or soda, the bottle will usually tell you how much the container holds in ml.

Key words
standard units
litres
millilitres
non-standard units

Mental maths

Work with a partner.

1 Say the first ten multiples of 1000 together.
2 Say the first ten multiples of 500 together.
3 Say the first ten multiples of 250 together.
4 500 ml is half a litre. 250 ml is one quarter of a litre. Say these millilitre amounts as litres. The first two have been done for you.

- a 1500 ml = 1 $\frac{1}{2}$ litres
- b 1250 ml = 1 $\frac{1}{4}$ litres
- c 3500 ml = ____ litres
- d 3250 ml = ____ litres
- e 3750 ml = ____ litres
- f 10 500 ml = ____ litres
- g 10 750 ml = ____ litres
- h 12 500 ml = ____ litres

1 Write each amount in litres.
- a 1000 ml
- b 2000 ml
- c 4000 ml
- d 10 000 ml
- e 1500 ml
- f 250 ml
- g 750 ml
- h 1750 ml

2 Write each amount in ml.
- a 0.5 ℓ
- b 0.25 ℓ
- c 0.1 ℓ
- d 0.4 ℓ
- e 1.2 ℓ
- f 3.8 ℓ
- g 12.7 ℓ
- h 20.3 ℓ

3 Copy and complete these rules.
- a When you convert capacity amounts from ____ to ____, you divide by 1000.
- b When you convert capacity amounts from ____ to ____, you multiply by 1000.

4 Write each set of capacities in order from smallest to greatest.
- a 3 litres   450 ml   4500 ml   0.8 ℓ   0.5 ℓ   1290 ml
- b 4.8 ℓ   4500 ml   50 ml   500 ml   1.5 ℓ

Litres and millilitres

5 a A cup is 250 ml. Copy and complete this table to show the relationship between cups and litres.

Cups	0		4				12
Litres	0	$\frac{1}{2}$		$1\frac{1}{2}$	2	$2\frac{1}{2}$	3

b Ato says that the table shows that to convert from cups to litres, you multiply by 2. What mistake has he made?

c Write the rules for converting from cups to litres and from litres to cups.

6 For baking and cooking, we may use measurements of less than a cup. In ml, how much is:

a $\frac{1}{2}$ cup?   **b** $\frac{1}{3}$ cup?

c $\frac{1}{4}$ cup?   **d** $\frac{3}{4}$ cup?

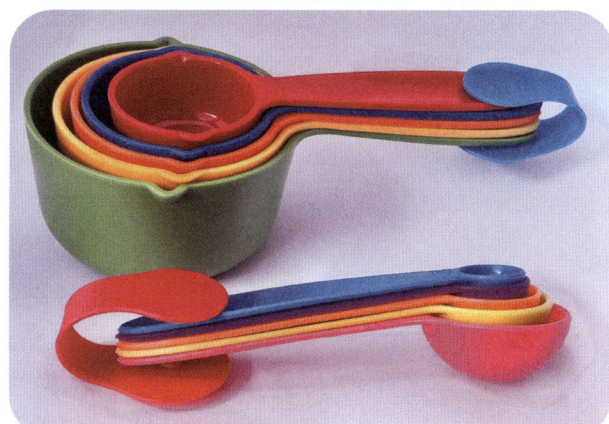

7 1 teaspoon is equal to 5 ml. Convert these measures to ml.

a 1 tablespoon = 3 teaspoons

b $\frac{1}{2}$ tablespoon

c $\frac{1}{2}$ teaspoon

8 Which is more: 3 tablespoons or $\frac{1}{8}$ cup?

Problem solving

1 Use containers that have a 1-litre, 2-litre, 5-litre and 10-litre capacity. You can use empty soda bottles, milk bottles and buckets. You will also need a container or measuring jug with a 500-ml capacity. Count how many 500-ml amounts each of the larger containers can hold.

Hint
Use plastic bottles or buckets from food products.

2 Look at the containers.

A 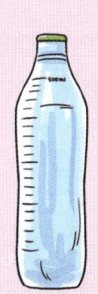 1 ℓ   B 500 ml   C 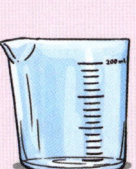 200 ml   D 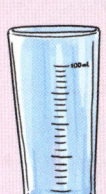 100 ml

a What is the capacity in millilitres of container A?
b What is the capacity of B and C together?
c How many of D would fill B?
d How many of C would fill A?
e Which of the containers could you fill twice to fill A?
f How much liquid would you need to fill all of the containers to half capacity?
g How much liquid would you use to fill all of the containers to their full capacity?

Section 6 Measurement Chapter 11 Capacity and volume

(continued)

3. Abigail's parents invite six guests for dinner. They cook rice and peas. They need $\frac{1}{2}$ cup of uncooked rice per person.

   Hint
   4 cups = 1 litre

   a How many cups of uncooked rice do they use?
   b To cook the rice, they need double the number of cups of water. How many cups of water do they need?
   c What is the correct amount of water in ml?
   d After it is cooked, the rice will double in volume. How many cups of cooked rice do they make?

4. A concentrate is a strongly flavoured liquid that you dilute with water. Here are two different concentrates.

Sorrel Juice

Recipe A:
Start with 3 cups of water.
Mix in 2 cups of concentrate.
Stir and enjoy!

Orange Juice

Recipe B:
You need 5 cups of water.
Add 3 cups of concentrate.
Mix, and your orange juice is ready!

   a Copy and complete the following sentences, keeping in mind that 1 cup is 250 ml.

   Recipe A requires ____ ml of concentrate and ___ ml of water. It makes ___ ml of juice altogether.

   Recipe B requires ___ ml of concentrate and ___ ml of water. It makes ___ ml of juice altogether.

   b How much of concentrate A would you need to make 3 litres of sorrel juice?
   c How much of concentrate B would you need to make 3 litres of orange juice?

5. A bathtub has a capacity of 150 litres.
   a If the bathtub is one third full, how much water does it hold?
   b If the tap pours out 5 ℓ every minute, how long will it take to fill the bath to full?

Investigating capacity and volume

Key maths idea

The volume of a solid is the amount of space it occupies. Capacity is the amount a container can hold. We can describe the volume of solid and hollow objects, but only hollow containers have capacity. You can think of capacity as how much liquid you can pour into a container.

Key words
conservation of volume

Think about what happens when you pour the same amount of water into different containers. The volume of water stays the same, even if the shape of the container is different. This is called the **conservation of volume**.

Solid objects also conserve their volume. This can help you to understand the relationship between capacity (measured in millilitres and litres) and volume (measured in cubic units).

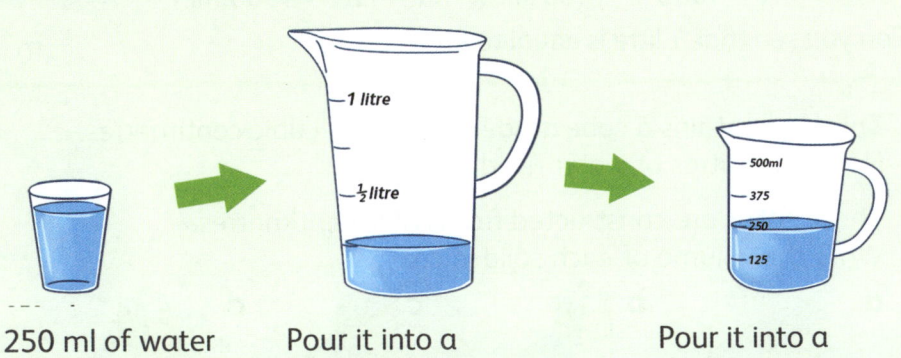

250 ml of water → Pour it into a 1-litre container → Pour it into a 500-ml container

Full STEAM ahead Exploring volume and capacity

1. Draw a table like the one below.

Container	Cubes (estimate)	Cubes (actual count)	Water (ml)	Difference between ml and number of cubes
A				
B				
C				

You will need:
- three plastic boxes, such as lunch boxes or other small containers, in different sizes
- cubic centimetre blocks
- water
- a 250 ml measuring jug marked in millilitres.

2. Label your containers A, B and C. Start with the smallest. Estimate how many cubic centimetre blocks you think will fit into it. Write your answer in the table.

3. a Use the blocks to check how many cubes fit inside. Write the answer.
   b Judge whether the answer seems reasonable in comparison to your estimate. If not, look at the estimate and the answer. Which was incorrect? Why?

4. Remove the blocks, and use the measuring jug to fill the container with water. Note in the table how many millilitres of water the container holds.

5. Follow steps **1** to **3** for all your containers.

6. Pour some water into the jug. Slowly put cubic centimetre blocks into the water. Watch the water level go up.
   a How many blocks do you need to add so that the water level goes up by 10 ml?
   b What does this tell us about the volume occupied by each block?

Section 6 Measurement Chapter 11 Capacity and volume

The relationship between volume and capacity

Key maths idea

In the 'Full STEAM ahead' investigation, you saw that 1 millilitre takes up 1 cm³.
A single row of ten cubic centimetres takes up 10 cm³.
Now imagine a square layer made up of ten rows of cubic centimetre blocks:
10 × 10 = 100   One layer occupies 100 cm³.
Now imagine a cube constructed of ten layers.
10 × 10 × 10 = 1000   You know that 1 litre = 1000 ml.
Can you see that 1 litre is equal to 1000 cm³?

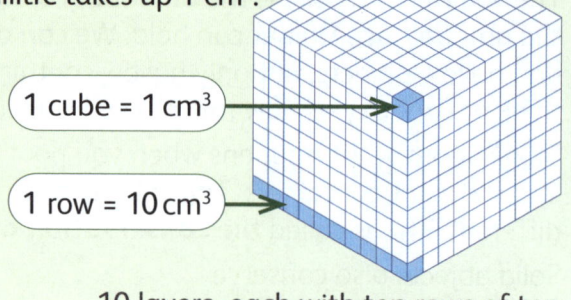

1 cube = 1 cm³
1 row = 10 cm³

10 layers, each with ten rows of ten

1   This jug contains a cube made up of 1000 cubic centimetres.
    How many litres of water filled the jug?

2   These solids are constructed from cubic centimetres.
    Write the volume of each solid in cm³.

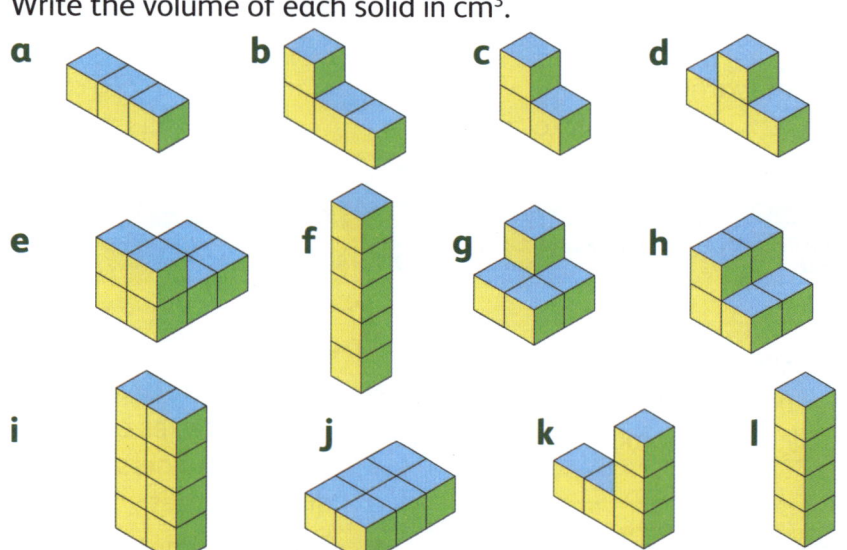

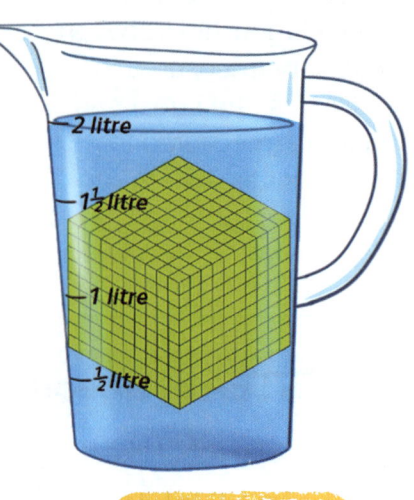

Hint
You will need to imagine the hidden blocks.

3   If you put solids **e**, **f** and **g** together into a measuring jug, how many ml of water must you add to make a total amount of 1 litre?

4   Omar has a set of solids with 100 of solid **h** and 20 of solid **e**. He puts them together into a box with a 1 litre capacity. What is the volume of the empty space left in the box?

What did you learn?

Look back at the work you did in this chapter. Rate your progress.
1 = I cannot do this. **2** = I need more practice. **3** = I understand it and feel confident.

Can you:
- explain the relationship between volume and capacity?
- describe how much a container can hold in millilitres or litres?
- describe the volume of an object in cubic units?
- solve problems involving volume and capacity?

Review: Capacity and volume

Key terms and concepts

1. a 1 litre = _____ millilitres
   b 250 ml = _____ litres
   c A cube that is 1 cm long, 1 cm wide and 1 cm deep is called a _____.
   d A cube that is 1 m long, 1 m wide and 1 m deep is called a _____.

2. Explain the relationship between litres and cubic centimetres.

Quick check

1. a $\frac{1}{2}$ litre = _____ millilitres
   b 0.3 litres = _____ millilitres
   c 0.05 litres = _____ millilitres
   d 800 ml = _____ litres
   e 250 ml = _____ litres
   f 25 ml = _____ litres

2. How much water would you need to fill containers with the following volumes? Give your answers in litres or millilitres.
   a 2000 cm^3
   b 3500 cm^3
   c 10 cm^3
   d 85 cm^3

3. A full container holds 25 cm^3 of sand. Tricia has an object with a volume of 16 cm^3. She wants to remove just enough sand to allow the object to fit. How many cm^3 of sand should she leave?

4. Look at this pattern for building stairs from construction blocks.

   How many blocks would you need to build a solid that has:
   a 5 steps?
   b 6 steps?

   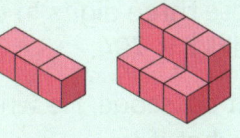
   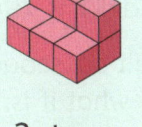

   1 step 2 steps 3 steps 4 steps

5. A shape is built out of cubic centimetre blocks. It has six layers, with 4 rows of 5 blocks per layer. What is its volume?

Challenge and investigate

1. Count the blocks of each shape to work out its volume.

   a
   b
   c
   d

2. A moving company provides three sizes of boxes. Size A is half size B and twice size C. A truck can fit 80 boxes of size C.
   a Write the box categories in order of size from smallest to greatest.
   b How many size B boxes can fit in the truck?
   c In a delivery, there are equal numbers of all three sizes. What is the greatest number of boxes that could fit?

3. Sasha has some small boxes like this.

   Each box is made up of 8 cubes.

   She wants to pack the smaller boxes into a big box like this:

   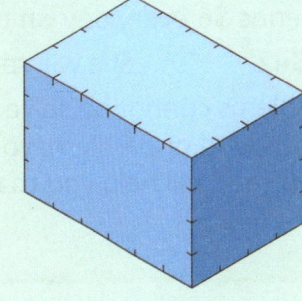

   a How many smaller boxes will she need to fill the bigger box?
   b Explain to a partner how you worked it out.

SECTION 7

Chapter 12 Solving real-life problems

In this chapter, you will:
- explore different strategies for solving routine and non-routine problems
- create and solve problems involving money and other real-life contexts
- solve problems involving unequal sharing.

Starting point

This is a combination-lock padlock. Each section of the lock has the digits 0 to 9 on it. You choose three digits as your passcode combination. The padlock will only unlock if you turn the dials to show those three digits in the correct order. The passcode on this padlock is 794.

1. Chin sets a three-digit passcode. He writes these clues so he can remember what it is:

   - It has three different odd digits.
   - If it was a decimal with two places, it would round to 5.8.
   - It can be divided by 3, but not by 9.

   Use the clues to work out the passcode.

2. How many passcodes are possible on this lock?
   Think carefully about how you could work this out. Share your ideas with your group.

Different types of problems

Key words
strategy
pattern

Key maths idea

Some problems can be solved using mathematical operations.

Example 1:
Nesha spends $6 every day on transport. How much will she spend in 7 days?
$6 × 7 = $42   She will spend 42 dollars.

Some problems cannot be solved by doing a mathematical operation. To solve this kind of problem, you have to think about it and find a **strategy** or **pattern** that will help you to work out the solution.

Hint
You can work out the answer mentally, but you must write the operation you used.

(continued)

Example 2:

Keisha has three counters: a red, a blue and a yellow.

How many ways are there for her to arrange the three counters in a row?

Mariah's strategy:

RBY   RYB
BYR   BRY
YRB   YBR

There are 6 ways to arrange the counters.

Billy's strategy:

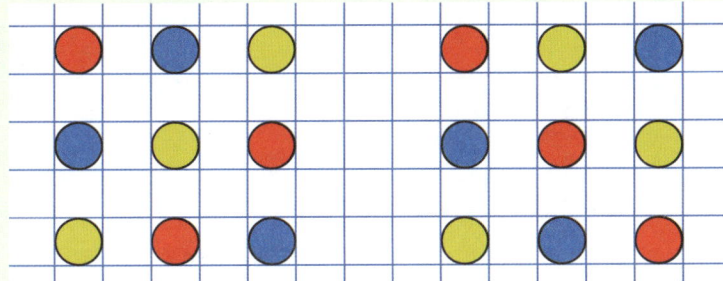

Mariah used initial letters to represent the colours. She made an organised list to find all the combinations.

Billy modelled the problem using counters.

Both strategies led to the correct solution.

1. How can you arrange six counters on this game board so that no three of them are in a line?

2. How many ways are there to make $20 using any combination of smaller bills?

3. Marcus had $230. He spent half of it on Friday. On Saturday, he spent $25. How much did he have left?

4. If there are 36 wheels, how many bicycles are there?

5. Mr Thomas has 96 oranges. He wants to pack them in bags of 6.
   a. How many bags will he need?
   b. How much money will he make if he sells each bag for $4?
   c. Would he make more money or less money if he packed the oranges into bags of four and sold each bag for $3?

Hint

Read each problem.
- Choose a strategy.
- Implement the plan or solve.
- Check the answer using another strategy or mental skill.

Talking maths

How do you know which operation to use to solve a problem? Tell your partner how you decide.

Section 7 Problem solving Chapter 12 Solving real-life problems

Problem-solving strategies

> **Key maths idea**
>
> ### Act it out or use concrete models
> The first step in problem solving is to read and understand the problem. Once you know what you need to find out, you can choose the strategy that you think is best.
>
> You can act out or use real objects to model simple problems that involve small numbers.
>
> **Example:**
>
> A taxi picks up passengers on the way to town. Three people get into the empty taxi at the first stop. At the second stop, one person gets out and two more people get in. At the third stop, three people get in and two people get out. How many people are in the taxi now?
>
>
>
> I used stones to model the problem and find the solution.
>
> We acted this out to find the solution.
>
> ### Guess, check and improve
> In this method, you start by making a sensible guess. Then you check whether or not your guess is correct. If not, you try again with an improved guess.
>
>
>
> This is a useful method for solving problems, but it can take a long time.
>
> **Example 1:**
>
> The sum of two numbers is 41 and the difference between them is 9. What are the numbers?
>
> | | A + B = 41 | A − B = 9 | |
> |---|---|---|---|
> | Guess 1: | 20 + 21 = 41 | 21 − 20 = 1 | Numbers must be further apart |
> | Guess 2: | 26 + 15 = 41 | 26 − 15 = 11 | Getting closer |
> | Guess 3: | 25 + 16 = 41 | 25 − 16 = 9 | Correct |
>
> The numbers are 25 and 16
>
> **Example 2:**
>
> A pilot has flown to 58 countries. Ten of these countries are in the Caribbean. The rest are in Africa and Europe. The pilot has flown to twice as many countries in Africa as in Europe. How many African countries has the pilot flown to?
>
>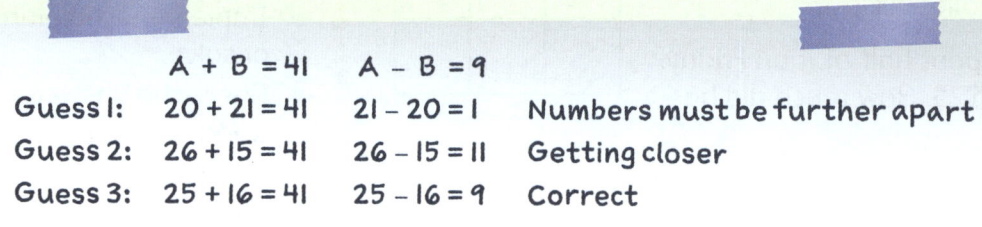
>
> The pilot has flown to 32 countries in Africa.

Problem-solving strategies

1. The digits 1 to 7 are written on seven cards.
   a. How many pairs of cards add up to 8?
   b. How many ways are there to make sets of three cards that add up to 10?

2. Josiah is 6 years older than Mia. The sum of their ages is 20. How old is Mia?

3. Five students are standing in line to go into the classroom.
   - Nigel is behind Gary. There is no one in between them.
   - Lisa is between Jemila and Gary.
   - Jemila is behind Derrick.

   Who is last in line?

4. Four students measured their height. Amar was taller than Keshon, but not as tall as Priya. Patricia was taller than Priya. Write the names of the students in order from shortest to tallest.

5. There are 20 vehicles in a parking lot. Half are white, one quarter are black, one tenth are gold and the rest are blue. How many vehicles of each colour are there?

6. Karl's age is $\frac{1}{5}$ of his dad's age. The sum of their ages is 54. How old is Karl's dad?

Key maths idea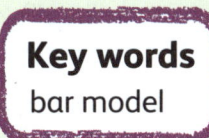

Draw a picture, diagram or bar model

Simple diagrams can help you to visualise a problem and work out what you need to do to solve it. Diagrams are very useful when the problem involves measurements or shapes.

Key words
bar model

Example 1:
A rectangular floor is 3 m wide. It is twice as long as it is wide. A rug (2 m wide and 4 m long) is placed on the floor. What area of the floor is not covered by the rug?

Draw a neat diagram of the room and the rug and label it using the information in the problem.

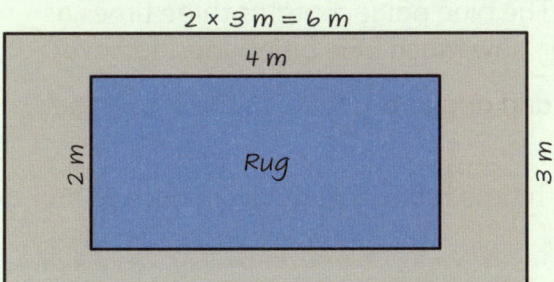

The diagram shows that the area of the floor minus the area of the rug is the area not covered.
Area of the floor = 3 m × 6 m = 18 m²
Area of the rug = 2 m × 4 m = 8 m²
Area not covered = 18 m² − 8 m² = 10 m²
10 m² of the floor is not covered by the rug.

Example 2:
Vanessa picked 120 limes. She sold $\frac{3}{4}$ of them at the market. On the way home from the market, she gave $\frac{1}{3}$ of the unsold limes to her friend. How many did she have left?

You can draw a **bar model** to show this problem.

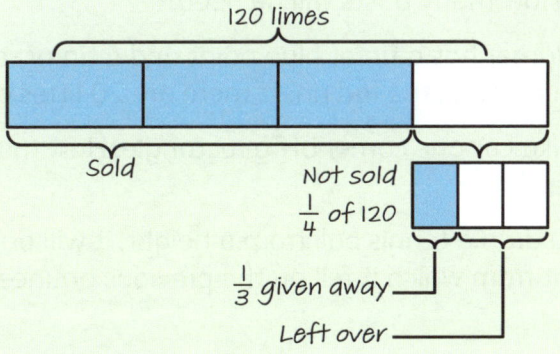

$\frac{1}{4}$ of 120 were not sold
$\frac{1}{4} \times 120 = 30$
$\frac{1}{3}$ of the leftover quarter was given away
$\frac{1}{3}$ of 30 = $\frac{1}{3} \times 30 = 10$ given away
30 − 10 = 20
Vanessa had 20 limes left over.

Section 7 Problem solving Chapter 12 Solving real-life problems

(continued)

Make an organised list or table

Making a list or table can help you to see patterns and make sure that you cover all the options.

Example 1:

A shop sells oil in 0.5 ℓ and 0.25 ℓ bottles. How many different ways are there to buy 1 litre of oil?

0.5 + 0.5	Two half-litre bottles
0.5 + 0.25 + 0.25	One half-litre and two quarter-litre bottles
0.25 + 0.25 + 0.25 + 0.25	Four quarter-litre bottles

There are 3 ways to buy 1 litre of oil.

Example 2:

Nikkita is 4 years old. Her brother Lee is three times her age. How old will Nikkita be when Lee is double her age?

Make a table showing their ages now. If Nikkita is 4, Lee is 12. Write both their ages for the next year and each year after that till Lee's age is double Nikkita's age.

Nikkita	4	5	6	7	8	9
Lee	12	13	14	15	16	

Nikkita will be 8 years old when Lee is double her age.

---

1 To open a locker, you press a letter from A to D and a number from 1 to 3. Oneika does not know the code for a locker, so she decides to start at A1 and work through the codes until she gets to the correct one. If the correct code is C1, how many combinations will she try before she gets the correct one?

2 How many ways are there to get a score of 8 or more when you roll two ordinary dice at the same time and add the dots to get your score?

3 A farmer wants to fence one side of a field that is 42 m long. The fence will have fence posts every 6 m. How many posts will he need?

4 Mr Warner has a tin of blue paint and a tin of red paint. The blue paint contains three times as much paint as the red tin. If there are 20 litres altogether, how much blue paint does he have?

5 Malaika cut one corner off a rectangle. How many sides and angles are there in the shape she has left?

6 If you drop a tennis ball from a height, it will bounce. Each time it bounces, it will bounce to half the height from which it fell on the previous bounce, like this:

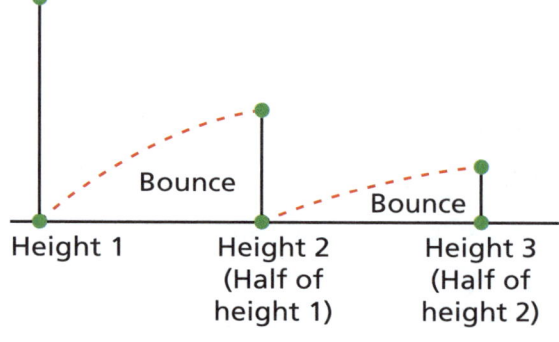

Tamia dropped a tennis ball. It bounced up 48 cm. How high will it bounce on its sixth bounce?

Problem-solving strategies

Key maths idea

Look for a pattern

Looking for patterns often goes together with tables.

Example:

Shereen is trying to get fitter by walking on a treadmill. She increases the distance she walks each day. The table shows how far she walked each day for 5 days.

Day	1	2	3	4	5
Distance (km)	1.3	1.4	1.6	1.9	2.3

If Shereen continues to increase the distance at the same rate, how far will she walk on day 10?

Look for a pattern in the distances.

Day	1	2	3	4	5
Distance (km)	1.3	1.4	1.6	1.9	2.3

+0.1   +0.2   +0.3   +0.4   +0.5 …

You can see that, each day, the distance increases 0.1 km more than it increased the previous day. You can use this pattern to work out the distance on day 10:

Day 6: 2.3 + 0.5 = 2.8   Day 7: 2.8 + 0.6 = 3.4

Day 8: 3.4 + 0.7 = 4.1   Day 9: 4.1 + 0.8 = 4.9

Day 10: 4.9 + 0.9 = 5.8   Shereen will walk 5.8 km on day 10.

Work backwards

You can solve some problems by starting with the end result and working back to find the solution.

Key words
inverse operations

Example:

Maggie plans to meet friends at the beach at 2.30 p.m. It takes her half an hour to walk to the beach. She wants to eat lunch and pack her beach gear before she leaves. It will take her 20 minutes to have lunch and 10 minutes to pack up her beach gear. At what time should she start eating lunch?

You can draw a number line to show the problem.

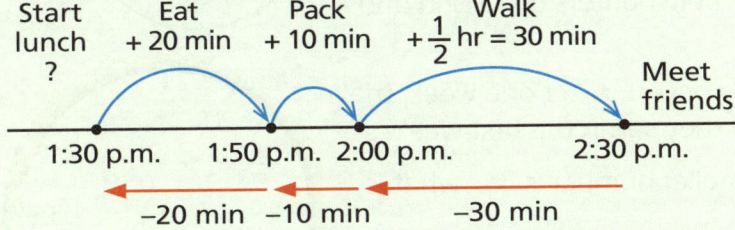

Work back from 2.30 p.m. using **inverse operations** to find the solution.

Maggie should start eating lunch at 1.30 p.m.

Section 7 Problem solving Chapter 12 Solving real-life problems

1. Mariah decides to sell coconuts at Queen's Park Savannah. On the first day, she makes $250 in sales. After that, she increases her sales by $40 a day. How much will she make in sales in 7 days?

2. Safraz starts with $256. He spends half his money each day.
   a. How much money will he have left after 3 days?
   b. How long will it take until he has 50c left?

3. A scientist is growing bacteria in a dish. Each day, the area covered by the bacteria doubles. If the bacteria takes 12 days to cover the area of the dish completely, how many days did it take to cover $\frac{1}{4}$ of the area?

4. Reza, Sharon, Patricia and Maleek collect and exchange stickers. When they have finished exchanging, Reza has 28 stickers. He had given 10 to Sam and received 12 from Patricia and 7 from Maleek. How many stickers did Reza have to start with?

5. A handyman charges $45 per hour, plus a callout fee of $35. If the final bill for a job was $327.50, how long did the job take?

Talking maths

What clues in a problem can help you decide which strategy to use? Share your ideas in small groups.

Hint

You will not be told which strategy to use to solve problems in your exam. You will need to read each problem carefully and decide which strategy you will use to solve it. There is often more than one strategy that you can use.

Problem solving

Solve these problems. You can use any strategy, but you must show all your working and remember to write the answers. If you are calculating, estimate before you start and use your estimate to check that your answer seems reasonable.

1. You can draw straight lines on a shape to divide it into regions, like this:

   Show how you can draw two straight lines on a clock face to make three regions with the sum of the numbers in each region the same.

2. There is one blue, one black, one yellow and one white counter. Ria takes the blue counter and Maggie takes the black counter. Keisha does not take the white counter. What colour counters do Keisha and Jenny take?

3. A class has 142 tickets to sell for a school concert. After one week, they still have 49 tickets to sell. How many did they sell in the first week?

4. The sum of two numbers is 7.26. If the smaller number is 3.4, what is the greater number?

5. A new hotel will have 35 rooms on the third floor. The rooms will be numbered from 301 to 335. The hotel uses wooden numerals to put the number on each door. They have 25 wooden 3s. How many more will they need to order to number all the doors?

6. Selina is working on a maths problem. She multiplies a number by 12 and then divides the result by 6, instead of multiplying by 6 and then dividing by 12. If Selina got 6 as her answer, what was the correct answer?

Profit and loss

Key maths idea

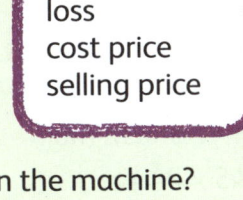

A business sells products or services. There is usually a difference between what it costs to make or buy a product, and the price that the business sells it for. If you buy or make something and then sell it for more than it cost you, you make a **profit**. If you sell it for less than it cost you, you make a **loss**.

You can calculate profit and loss if you know the **cost price** and the **selling price**.

Selling price − cost price = profit   Cost price − selling price = loss

Key words
profit
loss
cost price
selling price

Example 1:

A mechanic bought a machine for $1400. He sold it for $1260. What was his loss on the machine?

$1400 − $1260 = $140

He made a loss of $1400.

Example 2:

Use the information to work out the profit or loss on each sale.

a   Cost $700 to make   Sold for $630

The selling price is less than the cost price, so the difference is a loss.

700 − 630 = 70

Loss of $70

b   Sold for $7200   Cost $6520 to make

The selling price is more than the cost price, so the difference is a profit.

7200 − 6520 = 680

Profit of $680

1   Calculate the amount of profit or loss in dollars.
   a   Selling price $680, cost price $359
   b   Cost to make $18.50, sold for $25

2   Mona bought a second-hand phone for $530. She sold it online and made $75 profit. How much did she sell it for?

3   A farmer spent $4380 on equipment. He sold it later for $4150. How much was his loss?

4   Deborah bought one hundred caps to sell at the market. She paid $500 for the caps and then spent $2 per cap to have a logo printed on them. She sells the caps for $8 each.
   a   Did she make a profit or a loss?
   b   Calculate how much profit or loss she made altogether on the caps.
   c   Deborah makes $2.38 profit on each T-shirt. What is the cost price of the T-shirts?
   d   Deborah buys sweets in packs of 12. Each pack costs $4.00. How much profit does she make on each sweet?

Real-life maths

Business owners keep track of all their costs to work out what they spend. They compare this with the money the business earns to see whether it is making a profit. Think about a taxi driver. What costs do they have to consider to decide whether or not they are making a profit?

Section 7 **Problem solving** Chapter 12 Solving real-life problems

Savings

> **Key maths idea**
>
> **Saving** means keeping some of your money instead of spending it right away. You can put money into a **savings** account at the bank, so you can use it later for something important.
>
> **Key words**
> saving
> savings
>
> **Example**
>
> Arianna saves $45 every month. How much will she put into her savings in two years?
>
> 1 year = 12 months, so 2 years = 24 months
>
> 45 × 24 = ☐   Estimate: I know 12 × 5 = 60, so 12 × 50 = 600, so 24 × 50 = 1200
>
> ```
>     ¹
>    ²4 5
>  ×   2 4
>  ───────
>    1 8 0
>    9 0 0
>  ───────
>  1 0 8 0
> ```
>
> This is close to the estimate, so it looks reasonable.
> Arianna will put $1080 into her savings in two years.

1 Raj puts $13 per week into his savings account. How much will he put into the account in 20 weeks?

2 Stacy wants to save $1500 per year. She wants to save the same amount each month. How much should she save per month?

3 Amari has saved $468. How much more money will he need to save to have $800?

4 How many weeks will it take to save $120 if you put $7.50 a week into your account?

5 Sally saves $5 a week and Zahra saves $20 a month.
   a How much will each person have at the end of one year?
   b Why do they not have the same amount saved?

Hint
'Per' means 'each'. 'Per hour' means 'each hour' and 'per year' means 'each year'.

Problem solving

1 Zara puts $5 into a savings jar every day from Monday to Friday. If she has $42 in the jar on Friday, how much did she have on the previous Sunday?

2 Jabari works as a waiter. He works shifts on Mondays, Wednesdays and Fridays. He saves $15 per shift. If he has saved $105, how many weeks has he been saving for?

3 Ms Huggins inherits $5000. She decides to save $\frac{1}{4}$ of this amount. How much will she have left over?

4 Tyrone has saved $1250. He wants to buy a computer that costs $2099. How much more will he need to save to have enough money?

5 Vivek saves $45 a month. After $1\frac{1}{2}$ years, he withdraws half of his savings. How much will he have left in his bank account?

Salaries and wages

Key maths idea

Wages are the money paid to employees based on the number of hours they work. Wages are normally calculated and paid weekly.

Key words
wages
salary

Example 1:

Rishi earns $26 per hour. If he works 30 hours per week, what is his weekly wage?

$26 × 30 = ?   Estimate: 25 × 3 is 75, so 25 × 30 is 750

$26 = $20 + $6

$20 × 30 = $600 $6 × 30 = $180

$600 + $180 = $780   The answer seems reasonable. Check with a calculator.

A **salary** is a fixed amount of money paid to employees. Salaries are often worked out as an annual sum, but they are paid each month in equal amounts. Salaried employees receive the same salary each month regardless of how many hours they have worked.

Example 2:

Adara is offered a salary of $780 per month. What is her annual salary?

There are 12 months in a year.

$780 × 12 = ?   Estimate: 800 × 12 = 9600

×	700	80	Sum
10	7000	800	7800
2	1400	160	1560
			9360

The answer is lower than the estimate because the estimate was rounded up, but when the answer is checked with a calculator it is correct.

1. Alvin is paid a salary of $4280 per month. He receives an annual increase of $7800. What will his new monthly salary be?

2. Jeevan earns $26 per hour. If he works four shifts of 6 hours each during the week, what will his weekly wages be?

3. A cleaner works 21 hours per week and is paid $22 per hour. What is the cleaner's weekly wage?

4. Zaida is applying for a job. The annual salary is $46 200. How much will she earn per month?

Mental maths

1. For each card, make the largest and smallest possible salary amount using all the digits.

   a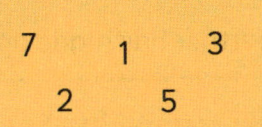
   b 1 1 7 2 6
   c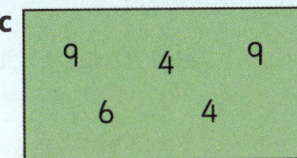

Section 7 **Problem solving** Chapter 12 Solving real-life problems

Full STEAM ahead   Create a problem pack

These four bar models represent different kinds of problems.

1 Copy each model onto a different card.

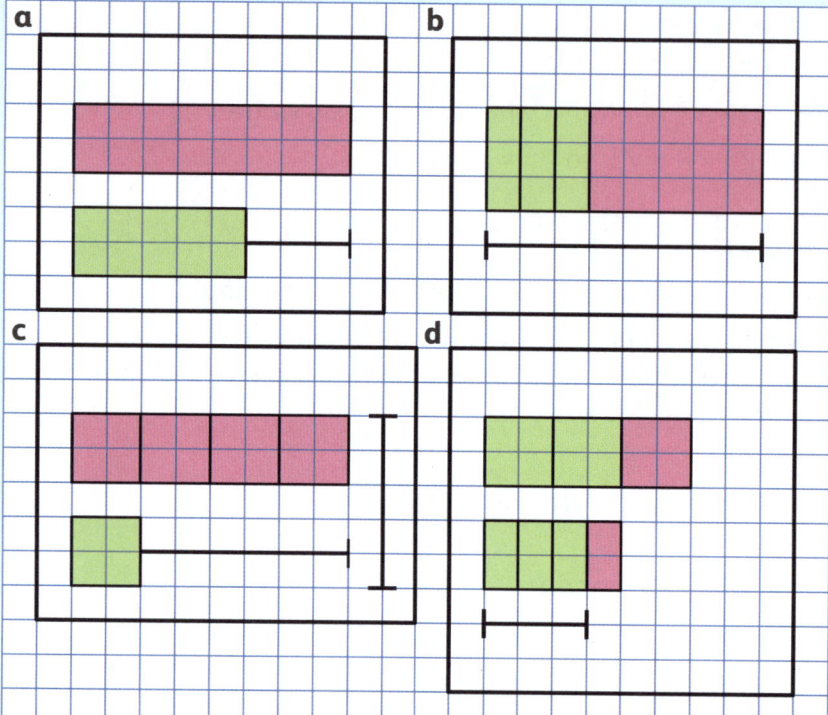

You will need:
- a ruler
- markers
- card or paper
- a calculator.

2 You are going to make up a word problem to match each model.

  a Think carefully about what the model shows you, and add information to it to match the problem.

  b Write the problem neatly onto the card. Here is an example:

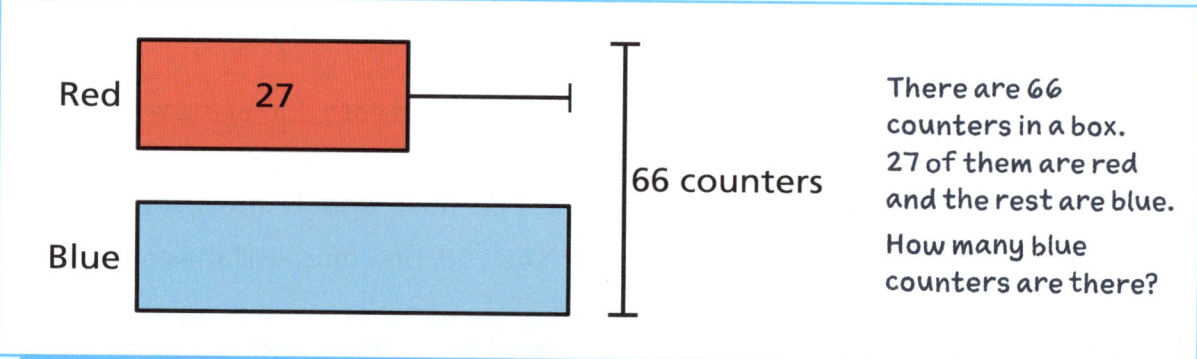

There are 66 counters in a box. 27 of them are red and the rest are blue. How many blue counters are there?

  c Work out the solution and write it onto the back of the card. Use a calculator to check that your solution is correct.

3 Exchange cards with a partner. Try to solve the problems that your partner has made up.

Talking maths

What did you find most challenging about creating your own problems? How did you overcome the challenges? Share your ideas in small groups.

Unequal sharing

Key maths idea

When you divide by a number, you are sharing the total equally. For example, 12 ÷ 3 gives you 4 equal groups of 3.

Sometimes, we share things unequally. For example, I share these oranges between two bowls so that one bowl has 4 more than the other. How many will be in each bowl?

Read how these students thought about and solved the problem.

Sasha

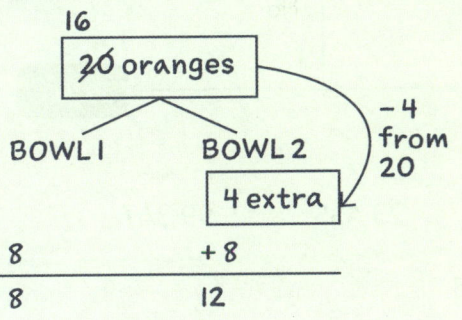

Jaden

Bowl A → ▲ oranges
Bowl B → ▲ + 4 oranges

▲ + ▲ + 4 = 20
▲ + ▲ = 16
▲ = 8

Both methods get the same result.
One bowl has 8 oranges and the other has 12.

1. Divide $45 between Randy and Diane so that Diane gets $5 more than Randy.

2. Share $250 dollars between Javid and Keisha so that Keisha gets $100 more than Javid.

3. Lisa has two lengths of rope. One piece of rope is 15 cm shorter than the other. The total length is 408 cm. How long is each piece?

4. Dillon paid $15.50 for a sandwich and a slice of cake. If the cake cost $3 more than the sandwich, what did the sandwich cost?

5. Over the weekend, 4125 people visit a market. If 523 more people visited on Sunday than on Saturday, how many people visited on each day?

Extension

6. Abigail, Dinesh and Jemila share 45 mangos. Abigail gets 5 more than Dinesh and Jemila gets 5 more than Abigail. How many do they each get?

Real-life maths

Jeevan says that a fair share is not always an equal share. Think up some real-life examples of where an unequal share is also a fair share.

Section 7 Problem solving Chapter 12 Solving real-life problems

Problem solving

1 Keshon made up this number-sorting game for five-digit numbers.

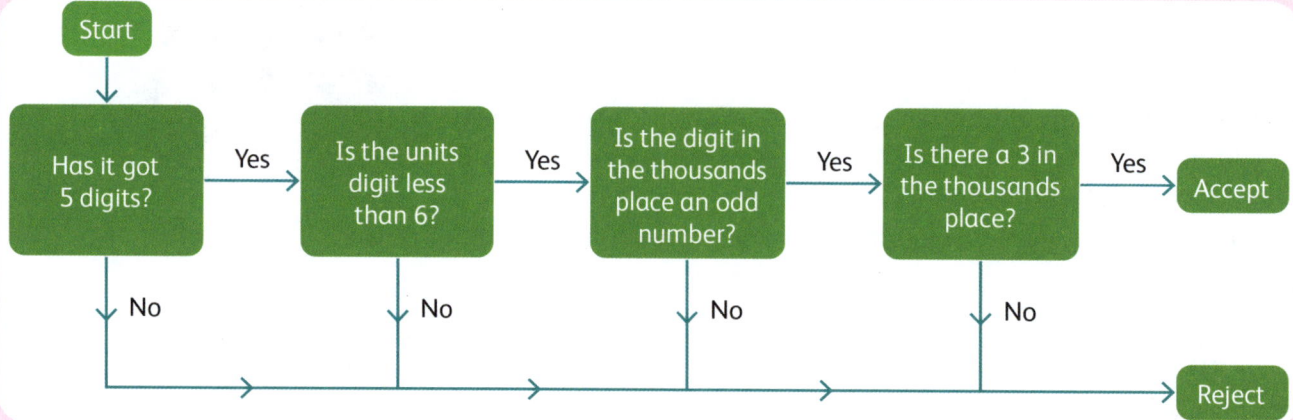

   a Which of these numbers will be rejected?

   5679   13 269   23 900   25 239   23 455   99 347

   b List five different five-digit numbers that will be accepted.

2 Karimah found a maths game online. The aim of the game is to arrange four dominoes in a square pattern so that the sum of the dots on each side of the square is the same.

The dominoes are arranged like this:

This set has sides that sum to 10:

Use a set of real dominoes if you have one. If not, your teacher will supply a printed set that you can cut up and use.

   a Find another way to arrange four dominoes to get sides equal to 10.
   b Make squares with sides of:

   4   5   6   7   8   9   16

Draw the squares that you make.

What did you learn?

Look back at the work you did in this chapter. Rate your progress.
1 = I cannot do this. 2 = I need more practice. 3 = I understand it and feel confident.

Can you:
- create and solve problems using a range of strategies?
- apply mental strategies to estimate?
- create single and multi-step problems?
- solve problems involving unequal sharing?

Review: Solving real-life problems

Key terms and concepts

1. Make a flow chart to show how to solve problems. Use some of these phrases:

   First …   Then …   Next …   When you have an answer …   Always …

2. Write a short definition of each word.
   - a profit
   - b loss
   - c wages
   - d salary
   - e savings

Quick check

1. Sadia has 18 counters. How many ways can she arrange them to form a rectangle?

2. A water tank can hold 10 000 litres. Ms Roberts's tank contained 4125 litres at the start of September. During the month, another 1476 litres flowed into the tank. How much more water can the tank hold before it is full?

3. A plant was 98 cm tall at the end of July. It increased $\frac{1}{2}$ cm in height during August and grew another $\frac{3}{4}$ cm taller in September. How tall was it at the end of September?

4. Calculate the profit or loss.
   - a Cost price $412, selling price $650
   - b Selling price $500, cost price $712

5. What would Marcus's wages be if he worked for 35 hours and earned $26 per hour?

6. A square has a perimeter of 1.2 m. What is the length of one side in centimetres?

Challenge and investigate

1. Josiah says 'I am thinking of a number. When it is multiplied by 3 and the product is increased by 5, the result is 17.' What is the number he is thinking of?

2. Nadia buys and sells paintings. She paid $420 for two paintings. She sold one for $348 and the other for $364. What was her profit on the two paintings?

3. The solution to a problem is: 1050 − 209 = 841. Write a word problem that fits this solution.

4. A rectangle is 3 times as long as it is wide. If its perimeter is 48 cm, what is its area?

5. A ship used 8760 litres of fuel in March and April. If it used 1000 litres more in April, how much did it use in March?

6. Malaika gets $15 per week for doing chores. She saves one third of her money each week. How long will it take her to save $50?

7. Ashanti bought a gift and a card for her mother. The gift cost $23 more than the card and she spent $35 in total. What was the cost of the card?

8. A carpenter plans to make 22 kitchen stools. Each stool can have 3 legs or 4 legs. He has 81 legs in stock and he wants to use them all. How many stools can he make?

SECTION 8

Chapter 13 Mass and weight

In this chapter, you will:
- use algebraic reasoning to find unknown mass or weight
- solve problems involving mass and weight.

Key words
balanced
mass
scales

Starting point

1. This scale has two red items on the left and three yellow items on the right.

    a. We can say that the scale is **balanced**. What does 'balanced' mean?

    b. What will happen if you add one more red item to the left-hand side? Why?

    c. What will happen if you remove one yellow item from the right-hand side? Why?

    d. What will happen if you add one yellow item to each side? Why?

    e. How could you describe the **mass** of the red items in informal units?

2. The **scales** on the left are weighing a watermelon. The scales on the right are weighing four avocadoes. Both sets of scales are balanced.

   a. What is the mass of the watermelon?   b. What is the mass of four avocados?

   c. What is half the mass of the watermelon?

   d. If the avocados are all the same mass, what is the mass of each one?

   e. How many avocados would you need to equal the mass of one watermelon?

Algebraic thinking

Key maths idea

When a scale is balanced, it means that the objects on both sides have the same mass. This scale is balanced. We do not know the mass of the square object, but we can use a letter to label it. We have used n.

Key word
equation

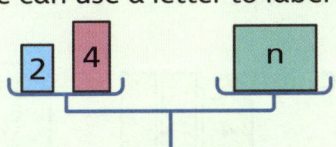

The square object has the same mass as the two weights in the other pan.
We can write this as 2 + 4 = ☐
We know that 2 + 4 = 6
So, ☐ = 6
The square has a mass of 6.

This scale below is also balanced. We have two identical squares with an unknown mass (y) and a 14-kg mass.
Each square must be equal to half of 14.
We can write this as $2 \times y = 14$
So, $y = 14 \div 2$
$y = 7$
We can also write this as $y + y = 14$.
We know that $7 + 7 = 14$, so $y = 7$.

Hint
You can use any letter or shape to represent an unknown amount.

Mental maths

1. Work out the total masses.
   a 3 kg + 4 kg + 8 kg
   b 200 g + 100 g + 50 g
   c 9 kg + $2\frac{1}{2}$ kg
   d 50 g + 50 g + 50 g + 50 g + 50 g
   e 2000 g + 400 g + 80 g
   f 8 kg + $2\frac{1}{2}$ kg + $1\frac{1}{4}$ kg
   g $\frac{1}{2}$ kg + $\frac{1}{2}$ kg + $\frac{3}{4}$ kg
   h $\frac{3}{4}$ kg + 2 kg + $\frac{1}{2}$ kg + $\frac{1}{4}$ kg

2. Natasha wants to make 450 gram packs of sweets. What mass of sweets will she need to add to each pack to make 450 grams?
   a 150 g
   b 280 g
   c 399 g
   d 110 g

1. Determine the unknown mass to balance each scale. Show your working.

   a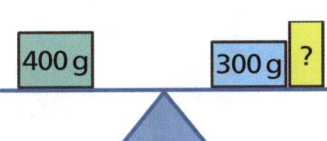
   b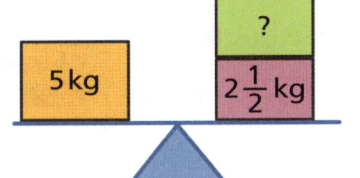
   c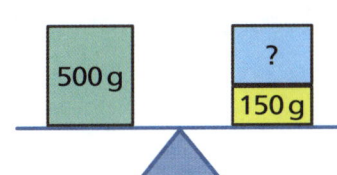

Section 8 Measurement Chapter 13 Mass and weight

d 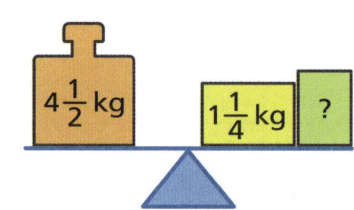   e 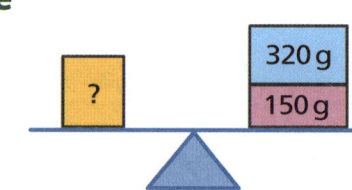   f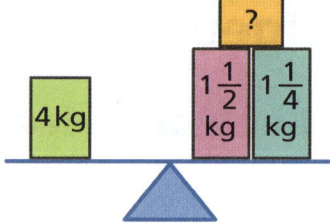

2 These scales are balanced. All the masses are in kilograms, and identical shapes have the same mass. For each balance, work out the mass of one shape.

a 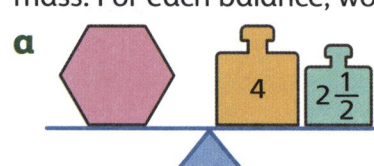   b   c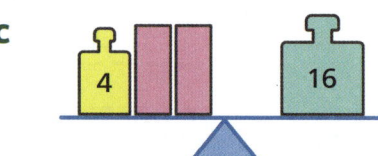

d 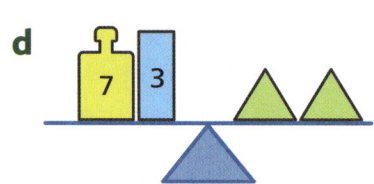   e 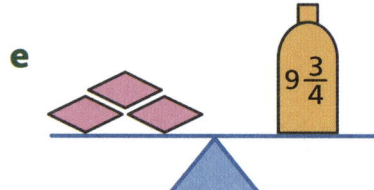   f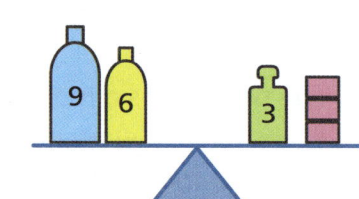

Talking maths

How could you balance this scale? Suggest two ways.

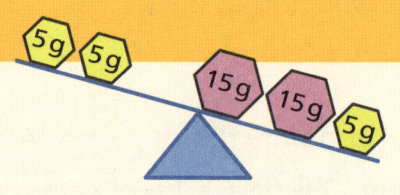

Full STEAM ahead Using a measuring scale

Oneika wanted to know how much a pea weighs. She tried two different kitchen scales.

She put the pea on the scales. They both showed no mass.

Sometimes small objects are so light that their mass does not show on a kitchen scale.

Oneika decided to use logical thinking to solve this problem. She counted out and weighed 20 peas. The scale showed a mass of 10 grams.

20 peas = 10 g

1 pea = 10 ÷ 20 = $\frac{1}{2}$ g

Use Oneika's ideas to find the mass of small objects.

Examples of objects you could try are paper clips, staples, seeds, small shells, pins or needles, sticky notes, leaves and plastic components.

Record your work in your exercise book.

You will need:
- a scale
- some small objects to weigh.

Balancing the scales

Key maths idea

When the scales are balanced, you can use an equal sign (=) to show the relationship between the objects on both sides of the scale.

These scales are balanced. The equation below each scale shows the relationship between the objects in the pans.

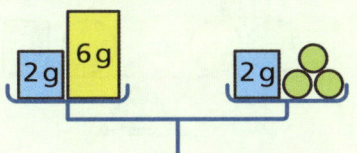
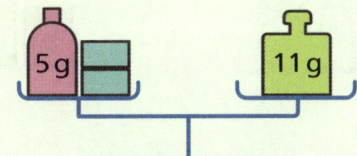

2 g + 6 g = 2 g + 3 circles
What is the mass of 1 circle?

5 g + 2 rectangles = 11 g
What is the mass of 1 rectangle?

You can move objects around to find the unknown masses.
- Whatever you do, keep the scales balanced.
- If you remove a mass from one side, you must remove the same mass from the other side.

Example 1:

You can remove 2 g from each pan to get the circles on their own on one side.

That leaves 6 g = 3 circles

6 ÷ 3 = 2

So, each circle weighs 2 g.

Example 2:

You can remove 5 g from each side to get the rectangles on their own.

Subtract the 5 g from the 11 g on the right to keep the balance.

11 g – 5 g = 6 g

That leaves 6 g = 2 rectangles

6 ÷ 2 = 3

So, each rectangle weighs 3 g.

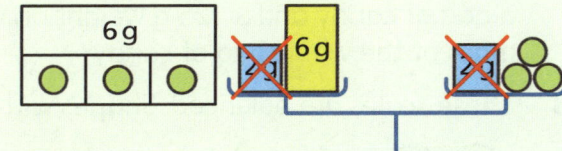

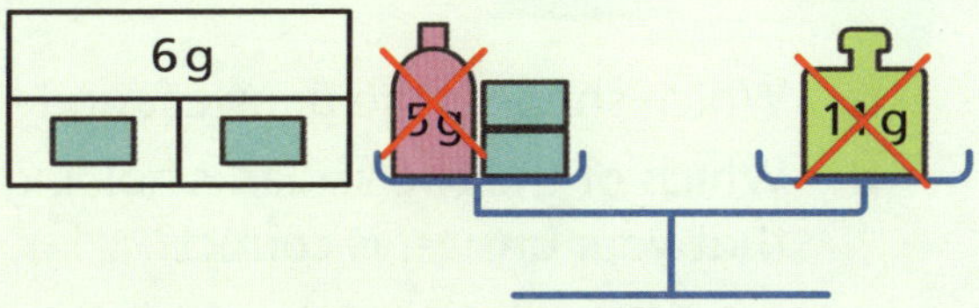

You can use these ideas to solve problems involving unknown masses without any numbers.

Example 3:

These scales are balanced, and shapes that are the same colour have the same mass. What could you add to the third scale to balance one blue triangle?

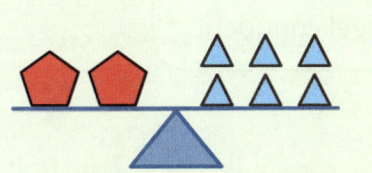
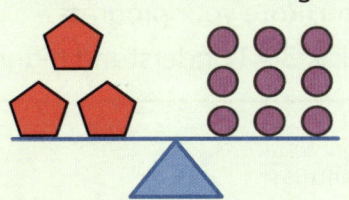

2 pentagons = 6 triangles   3 pentagons = 9 circles
1 pentagon = 3 triangles   1 pentagon = 3 circles
3 triangles = 3 circles, so 1 triangle = 1 circle
I would add one circle to balance the triangle on the third scale.

Section 8 Measurement Chapter 13 Mass and weight

1. For each balanced scale, work out the mass of one shape.

   a 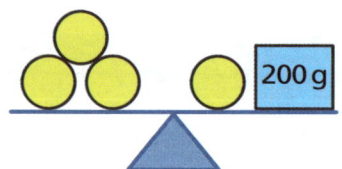   b    c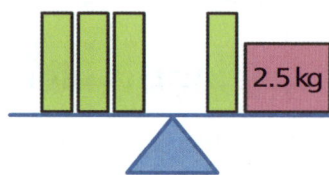

2. Determine the mass of one shape. Show your working.

   a 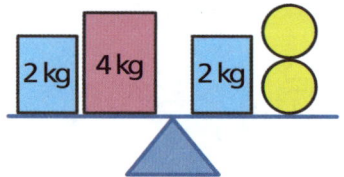   b 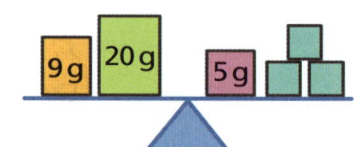   c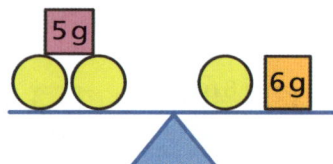

Problem solving

1. Lee has 5 g weights and 7 g weights. Show what weights he could use to balance each scale.

   a    b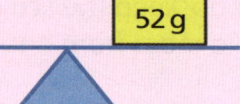

2. Priya is using a balance scale. She places a bag of candy in the pan on one side, and places half a bag of candy and a 125 g weight on the other side. If the scales are balanced, what is the mass of the whole bag of candy?

3. These scales are balanced. Shapes with the same letter have the same mass.

   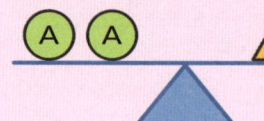   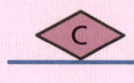 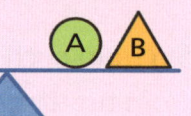

   a Which shape is the heaviest and which is the lightest? Justify your answer.
   b Which of the two scales is holding the greatest total mass? How can you convince someone that your answer is correct?

What did you learn?

Look back at the work you did in this chapter. Rate your progress.
1 = I cannot do this. **2** = I need more practice. **3** = I understand it and feel confident.

Can you:
- use algebraic thinking to find unknown values?
- solve problems involving mass and weight?

Review: Mass and weight

Key terms and concepts
Answer these questions to summarise what you learned in this chapter.
1. What does it mean if a scale is balanced?
2. If one pan is lower then the other, what does it tell you?
3. What is a mathematical name for a number sentence that contains an equals sign?
4. If two identical blocks + 3 kg = 10 kg, what is the mass of one block?
5. What are the rules for working out an unknown mass when there are different items on each side of the scale? Why is this important?

Quick check
1. Twelve identical marbles have a combined mass of 300 grams. What is the mass of one marble?
2. If three identical balls weigh 600 grams and two identical blocks weigh 150 grams, what is the combined mass of one ball and one block?
3. Sharon weighs 25 paper clips and finds they have a mass of 12.5 grams. What is the mass of each paper clip?
4. Show how you can find the mass of one shape in each diagram.

   a b c

Challenge and investigate
1. Estimate the mass in grams of one banana. Assume all the fruits are about the same mass.

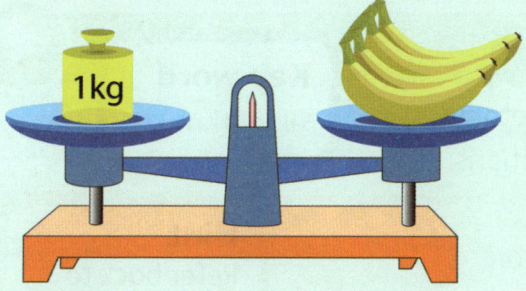

2. What is the mass of this watermelon?

3. Natasha weighs two boxes. Their total mass is 3.5 kg. If one box has a mass of 2.35 kg, what is the mass of the other box?
4. The average mass of 2 coconuts is 1.6 kg. Natasha adds a third coconut and the average mass becomes 1.4 kg. How heavy is the coconut that Natasha added?

SECTION 9

Chapter 14 Fractions 2

In this chapter, you will:
- add and subtract fractions and mixed numbers
- multiply fractions and mixed numbers
- divide fractions
- solve mixed problems involving fractions.

Key words
mixed number
improper fraction

Starting point

1. What number does each model represent? Give your answers as a **mixed number** and as an **improper fraction**.

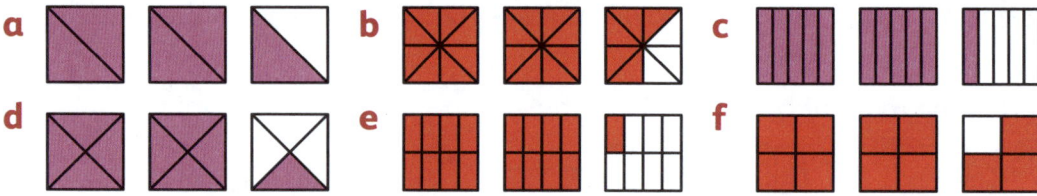

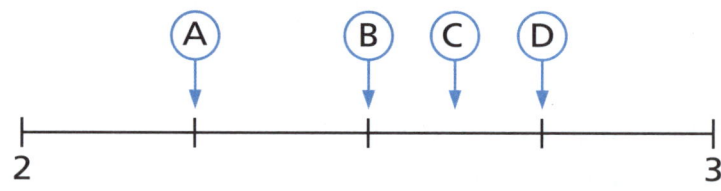

2. Four of the numbers represented in question 1 are marked on this number line. Which mixed number is shown at each position?

3. Draw your own number line to show the positions of the other two numbers.

Key maths idea

An **algorithm** is a set of instructions or rules for doing something. You already know some algorithms for calculating with fractions.

Key word
algorithm

1. What instructions do you follow to:
   a. add or subtract fractions with the same denominator?
   b. add or subtract fractions with different denominators?
   c. convert a mixed number to an improper fraction, and vice versa?
   d. simplify a fraction?

Hint
Refer back to Chapter 6 if you need a reminder of how to do these calculations.

Real-life maths

People sometimes use the word 'algorithm' when they talk about the way social media suggests videos and advertisements. Tell a partner what you understand about this kind of algorithm.

Adding and subtracting fractions

Key maths idea

Sometimes you need to add or subtract fractions with different denominators. Before you add or subtract, you must convert them to equivalent fractions with the same denominator, then you can just add the numerators.

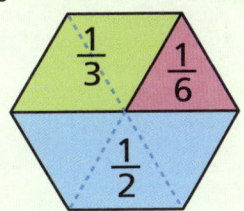

$\frac{1}{6} + \frac{1}{3} = \Box$

$\frac{1}{3} \times \frac{2}{2} = \frac{2}{6}$

$\frac{1}{6} + \frac{2}{6} = \frac{3}{6}$

In simplest form, this is $\frac{1}{2}$

So, $\frac{1}{6} + \frac{1}{3} = \frac{3}{6} = \frac{1}{2}$

Hint To simplify, divide the numerator and denominator by the same number: $\frac{3}{9} \div \frac{3}{3} = \frac{1}{3}$

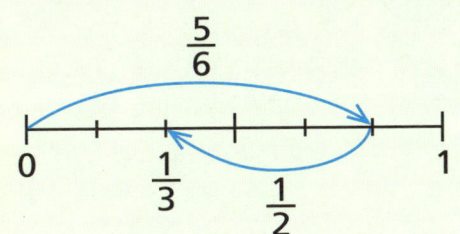

In simplest form, this is $\frac{1}{3}$

So, $\frac{5}{6} - \frac{1}{2} = \frac{2}{6} = \frac{1}{3}$

$\frac{5}{6} - \frac{1}{2} = \Box$

$\frac{1}{2} \times \frac{3}{3} = \frac{3}{6}$

$\frac{5}{6} - \frac{3}{6} = \frac{2}{6}$

1 Calculate:
- **a** $\frac{7}{10} + \frac{2}{10}$
- **b** $\frac{5}{8} + \frac{2}{8}$
- **c** $\frac{1}{6} + \frac{2}{6}$
- **d** $\frac{2}{10} + \frac{4}{10}$
- **e** $\frac{7}{8} - \frac{6}{8}$
- **f** $\frac{5}{6} - \frac{2}{6}$
- **g** $\frac{12}{12} - \frac{5}{12}$
- **h** $\frac{8}{9} - \frac{5}{9}$

Hint If your answer is an improper fraction, give it in its simplest form.

2 Calculate. Show your working. Simplify your answer if necessary.
- **a** $\frac{7}{10} + \frac{1}{5}$
- **b** $\frac{1}{10} + \frac{4}{5}$
- **c** $\frac{1}{2} + \frac{3}{10}$
- **d** $\frac{1}{4} + \frac{1}{8}$
- **e** $\frac{1}{4} + \frac{1}{2}$
- **f** $\frac{2}{5} + \frac{9}{10}$
- **g** $\frac{2}{3} + \frac{5}{12}$
- **h** $\frac{1}{3} + \frac{1}{9}$

3 Calculate. Show your working. Simplify your answer if necessary.
- **a** $\frac{9}{10} - \frac{1}{2}$
- **b** $\frac{11}{12} - \frac{1}{4}$
- **c** $\frac{7}{8} - \frac{3}{4}$
- **d** $\frac{19}{20} - \frac{1}{4}$
- **e** $\frac{1}{4} - \frac{1}{8}$
- **f** $\frac{4}{9} - \frac{1}{3}$
- **g** $\frac{2}{5} - \frac{3}{10}$
- **h** $\frac{7}{8} - \frac{1}{16}$

Adding and subtracting with mixed numbers

Key maths idea

There are different ways to add and subtract mixed numbers.
You can add or subtract the whole numbers and fractions separately.

$1\frac{2}{3} + 3\frac{2}{3} = \Box$

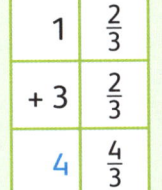 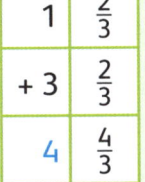  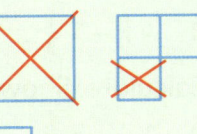

$2\frac{3}{4} - 1\frac{1}{4} = \Box$

$4 + \frac{4}{3} = 4 + 1\frac{1}{3} = 5\frac{1}{3}$ $1\frac{2}{3} + 3\frac{2}{3} = 5\frac{1}{3}$ $2\frac{3}{4} - 1\frac{1}{4} = 1\frac{1}{2}$

Section 9 Number Chapter 14 Fractions 2

(continued)

When the fraction parts of the mixed number have different denominators, use equivalent fractions to write them with the same denominator.

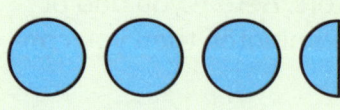

$3\frac{1}{2} + 2\frac{1}{4} = \square$

Convert into equivalent fractions

$\frac{1}{2} \times \frac{2}{2} = \frac{2}{4}$

Complete the calculation

$= 3 + 2 + \frac{2}{4} + \frac{1}{4}$

$= 5 + \frac{3}{4} = 5\frac{3}{4}$

You may need to regroup the whole-number part of a mixed number when you subtract. Look at the diagram to see how you can subtract $\frac{7}{8}$ from $3\frac{1}{8}$.

$3\frac{1}{8} - \frac{7}{8} = \square$

$2\frac{9}{8} - \frac{7}{8} = 2\frac{2}{8} = 2\frac{1}{4}$

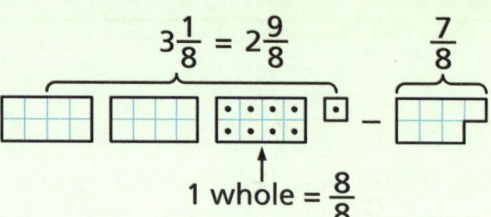

$3\frac{1}{8} = 2\frac{9}{8}$

1 whole = $\frac{8}{8}$

Hint
Remember to simplify the fraction part of the answer if possible.

1 Add:
 a $2\frac{2}{5} + 4\frac{1}{5}$
 b $3\frac{1}{2} + 4\frac{1}{2}$
 c $9\frac{1}{3} + 2\frac{1}{3}$
 d $4\frac{1}{10} + 2\frac{7}{10}$
 e $3\frac{2}{5} + 1\frac{3}{5}$
 f $2\frac{2}{3} + 1\frac{2}{3}$
 g $1\frac{3}{4} + 2\frac{3}{4}$
 h $3\frac{2}{5} + 1\frac{4}{5}$

2 Subtract:
 a $4\frac{3}{4} - 1\frac{1}{4}$
 b $3\frac{4}{5} - 1\frac{2}{5}$
 c $4\frac{9}{10} - 1\frac{8}{10}$
 d $5\frac{3}{7} - 2\frac{1}{7}$
 e $5\frac{3}{4} - 2\frac{3}{4}$
 f $7\frac{8}{10} - 4\frac{3}{10}$
 g $5\frac{3}{5} - 2\frac{4}{5}$
 h $7\frac{1}{4} - 3\frac{3}{4}$

3 Add. Show your working.
 a $3\frac{1}{2} + 2\frac{7}{10}$
 b $3\frac{3}{4} + 1\frac{1}{2}$
 c $2\frac{1}{3} + 2\frac{1}{6}$
 d $2\frac{1}{4} + 1\frac{7}{12}$
 e $2\frac{7}{10} + 1\frac{1}{5}$
 f $2\frac{1}{4} + 5\frac{1}{8}$
 g $1\frac{1}{2} + 3\frac{3}{8}$
 h $1\frac{2}{5} + 2\frac{3}{10}$

4 Subtract. Show your working.
 a $4\frac{3}{10} - 2\frac{7}{10}$
 b $3\frac{1}{8} - 1\frac{3}{8}$
 c $5\frac{1}{10} - 2\frac{3}{5}$
 d $5\frac{2}{3} - 1\frac{1}{6}$
 e $4\frac{4}{9} - 1\frac{1}{3}$
 f $6\frac{3}{5} - 1\frac{1}{10}$
 g $2\frac{3}{20} - \frac{19}{20}$
 h $4\frac{1}{2} - \frac{7}{8}$

5 Calculate. Show all the steps in your working.
 a $1\frac{1}{2} + 3\frac{1}{8} - 1\frac{3}{4}$
 b $3\frac{3}{4} - 1\frac{1}{8} + 1\frac{1}{2}$
 c $3\frac{4}{9} - 2\frac{1}{3} + \frac{5}{6}$
 d $2\frac{1}{3} + 4\frac{1}{4} - 2\frac{3}{4}$
 e $1\frac{5}{12} + 2\frac{1}{3} - 1\frac{1}{2}$
 f $3\frac{1}{2} + 2\frac{3}{5} - 3\frac{9}{10}$

Adding and subtracting fractions

Talking maths

These students have worked differently to add mixed numbers.

Sadia's method
$3\frac{1}{2} + 4\frac{3}{4}$
$= 3 + 4 + \frac{1}{2} + \frac{3}{4}$
$= 7 + \frac{2}{4} + \frac{3}{4}$
$= 7\frac{5}{4} = 7 + 1\frac{1}{4} = 8\frac{1}{4}$

Rishi's method
$3\frac{1}{2} + 4\frac{3}{4}$
$= \frac{7}{2} + \frac{19}{4}$
$= \frac{14}{4} + \frac{19}{4} = \frac{33}{4} + 8\frac{1}{4}$

What has each student done?

Which method makes most sense to you? Why?

How could you use Rishi's method to subtract $2\frac{1}{4}$ from $5\frac{3}{4}$? Tell your partner.

Key maths idea

You can regroup mixed numbers to write them as improper fractions before you add or subtract.

$4\frac{4}{5} - 2\frac{1}{10}$ Convert the answer back to a mixed number.

$= \frac{24}{5} - \frac{21}{10}$ Regroup mixed numbers to make equivalent improper fractions.

$= \frac{48}{10} - \frac{21}{10}$ Convert to equivalent fractions with the same denominator.

$= \frac{27}{10}$

$= 2\frac{7}{10}$ Convert the improper fraction to a mixed number.

Problem solving

1. Work out the missing value in each number sentence.
   a ☐ $- 4\frac{1}{2} = 1\frac{1}{6}$ b $2\frac{1}{4} +$ ☐ $= 4\frac{19}{20}$ c ☐ $- 3\frac{1}{4} = 4\frac{3}{5}$

2. Ms Francis is baking a cake.
   a She mixes $3\frac{1}{2}$ cups of flour with $1\frac{1}{3}$ cups of butter. How many cups is that altogether?
   b She has $4\frac{3}{5}$ cups of sugar. She needs $2\frac{4}{5}$ cups for the cake and $2\frac{1}{5}$ cups for the icing. Does she have enough sugar?
   c Ms Francis pours $2\frac{1}{3}$ litres of milk into a jug with a capacity of $5\frac{8}{12}$ ℓ. How much more milk can the jug hold?

3. Meela ran $12\frac{1}{10}$ km. Keisha ran $14\frac{3}{5}$ km. How much further did Keisha run?

4. What is the perimeter of a rectangular lawn $12\frac{2}{5}$ m long and $3\frac{7}{10}$ m wide?

5. A roti shop orders three bags of flour. Each bag weighs $4\frac{2}{5}$ kg. At the end of a week, one bag is empty, another has $2\frac{3}{5}$ kg in it and the third bag has $\frac{9}{10}$ kg. How much of the flour has the shop used?

Section 9 Number Chapter 14 Fractions 2

Multiplying with fractions

> **Key maths idea**
>
> You have already learned how to multiply fractions and whole numbers and how to find a fraction of an amount.
>
> **Example 1**
> What is $3 \times \frac{3}{8}$?
>
>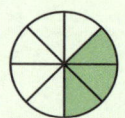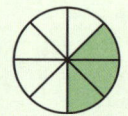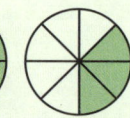
>
> We can find this using repeated addition.
> $\frac{3}{8} + \frac{3}{8} + \frac{3}{8} = \frac{9}{8}$
> So, three lots of three eighths is 9 eighths.
> We can get the same result by multiplying.
> $3 \times \frac{3}{8} = \frac{3}{1} \times \frac{3}{8} = \frac{9}{8}$
> Write 3 as a fraction with a denominator of 1
>
> **Example 2**
> Find $\frac{2}{3}$ of 15.
>
>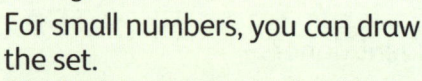
>
> For small numbers, you can draw the set.
> Divide the set into thirds. You can see that there are 5 stars in one third, so there are 10 stars in two thirds.
> In mathematics, the word 'of' tells you to multiply, so you can get the same result by multiplying.
>
> $\frac{2}{3}$ of $15 = \frac{2}{3} \times 15$
> $= \frac{2}{3} \times \frac{15}{1}$
> $= \frac{30}{3}$ Divide the numerator
> $= \frac{10}{1} = 10$ and denominator by 3 to simplify the fraction

1 Calculate. Write the answers as proper fractions in simplest form or as mixed numbers.
  a $4 \times \frac{2}{10}$ **b** $3 \times \frac{1}{4}$ **c** $9 \times \frac{1}{10}$ **d** $4 \times \frac{1}{8}$
  e $3 \times \frac{5}{10}$ **f** $4 \times \frac{1}{3}$ **g** $4 \times \frac{2}{5}$ **h** $9 \times \frac{1}{2}$

2 What is:
  a $\frac{2}{3}$ of 24?   **b** $\frac{1}{8}$ of 32?   **c** $\frac{4}{9}$ of 36?   **d** $\frac{2}{3}$ of 30?
  e $\frac{4}{5}$ of 100?   **f** $\frac{9}{10}$ of 20?   **g** $\frac{3}{4}$ of 60?   **h** $\frac{1}{2}$ of $\frac{3}{4}$?

3 Maggie swims $\frac{3}{5}$ of a kilometre every day for 5 days. How far does she swim in total?

4 How much water is in a 200 ℓ tank if it is $\frac{3}{5}$ full?

A fraction of a fraction

> **Key maths idea**
>
> To multiply fractions by whole numbers, you wrote the whole number as a fraction. Then you multiplied numerators by numerators and denominators by denominators.
> You can use this method to multiply fractions by other fractions.
>
> **Example 1**
> What is $\frac{3}{4}$ of $\frac{1}{2}$?
>
>
>
> 1 whole $\frac{1}{2}$ $\frac{3}{4}$ of $\frac{1}{2}$ Which is $\frac{3}{8}$ of the whole
>
> $\frac{3}{4}$ of $\frac{1}{2} = \frac{3}{4} \times \frac{1}{2}$
> $= \frac{3 \times 1}{4 \times 2}$
> $= \frac{3}{8}$
>
> The diagram shows that this is correct.

A fraction of a fraction

(continued)

Example 2

What is $\frac{2}{3}$ of $\frac{3}{5}$?

Think of $\frac{3}{5}$ as three equal parts out of five.

You can divide each fifth into thirds.

That gives you fifteen equal parts: fifteenths.

$\frac{2}{3}$ of each of the three fifths is shaded.

We multiply to find $\frac{2}{3}$ of $\frac{3}{5}$

$\frac{2}{3} \times \frac{3}{5}$

$= \frac{2 \times 3}{3 \times 5} = \frac{6}{15}$

$\frac{2}{3}$ of each fifth

6 parts out of 15

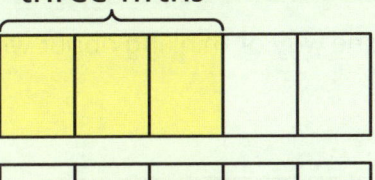
three fifths

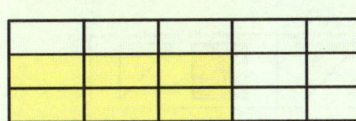

1 Calculate. Show your working.

a $\frac{1}{5} \times \frac{1}{3}$ b $\frac{1}{4} \times \frac{1}{6}$ c $\frac{1}{5} \times \frac{1}{9}$ d $\frac{1}{3} \times \frac{2}{3}$

e $\frac{2}{3} \times \frac{4}{5}$ f $\frac{3}{4} \times \frac{2}{7}$ g $\frac{3}{10} \times \frac{5}{4}$ h $\frac{3}{7} \times \frac{2}{3}$

i $\frac{3}{2} \times \frac{1}{4}$ j $\frac{2}{3} \times \frac{4}{5}$ k $\frac{3}{5} \times \frac{7}{11}$ l $\frac{3}{8} \times \frac{2}{5}$

Hint
When you are asked to simplify, it means do the operation and give the answer in its simplest form.

2 Simplify each of the following number sentences.

a $\frac{4}{5} \times \frac{1}{2}$ b $\frac{3}{7} \times \frac{2}{5}$ c $\frac{1}{3} \times \frac{2}{4}$ d $\frac{9}{10} \times \frac{1}{2}$

e $\frac{1}{3}$ of $\frac{1}{3}$ f $\frac{2}{7}$ of $\frac{3}{9}$ g $\frac{9}{10}$ of $\frac{2}{3}$ h $\frac{6}{10} \times \frac{5}{6}$

i $\frac{2}{3}$ of $\frac{4}{7}$ j $\frac{4}{5}$ of $\frac{2}{10}$ k $\frac{3}{4}$ of $\frac{4}{5}$ l $\frac{4}{3}$ of $\frac{2}{6}$

3 Each of these statements is false. Show why.

a $\frac{12}{15}$ is four times $\frac{3}{5}$ b $\frac{9}{10}$ is twice $4\frac{1}{2}$

c $\frac{4}{20}$ is twice $\frac{2}{5}$ d $\frac{20}{30}$ is equivalent to five times $\frac{4}{6}$

Problem solving

Work step by step to solve question 3. Show your working for each step.

1 Oneika bought 3 kilograms of mixed nuts. $\frac{3}{8}$ of them were pecans. How many kilograms of pecans were there?

2 Jaden eats $\frac{9}{10}$ of a packet of cereal each week. Kimani eats $\frac{1}{2}$ as much cereal as Jaden. How much of a packet does Kimani eat?

3 Sasha has $\frac{3}{5}$ of a bag of flour. She uses $\frac{1}{4}$ of the flour to make doubles and then uses $\frac{1}{2}$ of what is left to make roti. How much of the bag is left over?

4 Naresh feeds his two cats dry food. Slinkycat eats $\frac{3}{8}$ of a cup of food per day and Bulkycat eats $\frac{1}{2}$ cup of food per day. How many cups of food do the cats eat altogether each week?

Section 9 Number Chapter 14 Fractions 2

Multiplying with mixed numbers

Key maths idea

The diagrams show one way of thinking about what it means to multiply with mixed numbers.

Example 1
What is $3 \times 1\tfrac{1}{2}$?

We can find this by adding: $1\tfrac{1}{2} + 1\tfrac{1}{2} + 1\tfrac{1}{2} = 4\tfrac{1}{2}$

 $4\tfrac{1}{2}$

We can also multiply.

$1\tfrac{1}{2} = \tfrac{3}{2}$ (three halves)

$3 \times \tfrac{3}{2} = \tfrac{3}{1} \times \tfrac{3}{2} = \tfrac{9}{2}$

$\tfrac{9}{2} = 4\tfrac{1}{2}$

Example 2
What is the area of a rug that is $2\tfrac{3}{4}$ m long and $1\tfrac{1}{2}$ m wide?

	2	$\tfrac{3}{4}$
1	2	$\tfrac{3}{4}$
$\tfrac{1}{2}$	1	$\tfrac{3}{8}$
	3	$\tfrac{3}{4} + \tfrac{3}{8}$

Think of the rug as a grid.
Work out the area of each block in the grid by multiplying.
Add the totals for each column.
Write the answer as a mixed number.

$= \tfrac{6}{8} + \tfrac{3}{8}$
$= \tfrac{9}{8}$
$= 1\tfrac{1}{8}$

$3 + 1\tfrac{1}{8} = 4\tfrac{1}{8}$

You can also convert the mixed numbers to improper fractions and find the answer by multiplying.

$1\tfrac{1}{2} \times 2\tfrac{3}{4} = \tfrac{3}{2} \times \tfrac{11}{4} = \tfrac{33}{8} = 4\tfrac{1}{8}$

Talking maths

Explain why $\tfrac{7}{8} \times 2\tfrac{1}{2}$ will give a result less than $2\tfrac{1}{2}$ but $\tfrac{8}{7} \times 2\tfrac{1}{2}$ will give a result greater than $2\tfrac{1}{2}$.

1 Calculate mentally. Write the answers only.
   a $2 \times 3\tfrac{1}{2}$
   b $3 \times 1\tfrac{1}{4}$
   c $2 \times 3\tfrac{1}{5}$
   d $4 \times 10\tfrac{1}{4}$

2 Find the total mass of:
   a 7 boxes each weighing $3\tfrac{1}{3}$ kg
   b 5 boxes each weighing $1\tfrac{3}{4}$ kg

3 A small tin of paint holds $1\tfrac{1}{4}$ litres and a large tin holds $2\tfrac{1}{2}$ litres. What is the total volume of:
   a 3 small tins?
   b $2\tfrac{1}{2}$ small tins?
   c 7 large tins?
   d $3\tfrac{1}{2}$ large tins?
   e $\tfrac{1}{4}$ of a large tin?
   f $\tfrac{7}{8}$ of a small tin?

4 Rakesh runs laps round an oval track. The distance around the track is $1\tfrac{1}{8}$ km. Work out how far Rakesh runs each day.
   a Monday 2 laps
   b Tuesday $2\tfrac{1}{4}$ laps
   c Wednesday $2\tfrac{5}{8}$ laps
   d Thursday $\tfrac{9}{10}$ of a lap
   e Friday $1\tfrac{1}{2}$ laps
   f Saturday $\tfrac{11}{12}$ of a lap

Problem solving

1. Erica is in a steelpan group. Before a competition, they practise for $2\frac{3}{4}$ hours three times a week. How much time do they spend practising each week?

2. Tyrone jogs at a speed of $7\frac{3}{4}$ km per hour. How far will he run in $2\frac{1}{2}$ hours at this speed?

3. A dripping tap leaks $6\frac{1}{4}$ litres of water every hour. Ria puts an empty bucket under the dripping tap. After $3\frac{1}{2}$ hours, how many litres of water has the bucket collected?

4. Mr Ali has one flower bed that is $3\frac{1}{2}$ m long and $2\frac{3}{4}$ m wide and another flower bed that is $1\frac{1}{2}$ times larger. What is the total area of the two flower beds?

Full STEAM ahead   Designing with fractions of fractions

Tara designs and makes tiles. Read what she says about one of her designs.

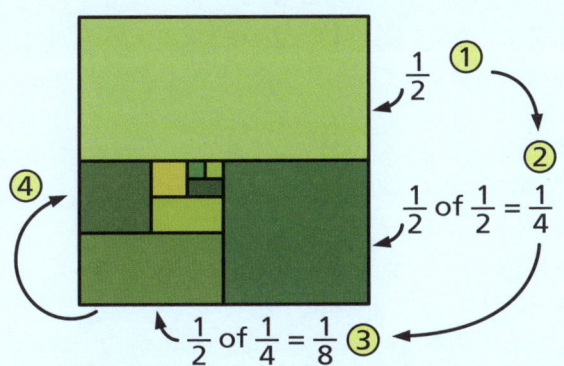

I use fractions in my designs. For this tile, I use halves.

I start by dividing the square into two halves.

Then I find half of one half. That is a quarter, by the way!

Next I find half of the quarter.

I carry on finding half of the remaining area until I get to the smallest area that I can paint.

Draw three squares with side lengths of 12 cm. Use squared paper if you have it.

Complete these two designs for three quarters and one third, in the same way Tara did for halves. The first fraction has been drawn for each one.

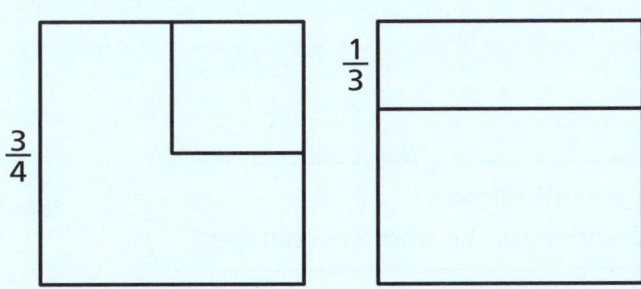

Choose your own fraction for the last design.
Colour your tiles and make a class display.

Section 9 **Number** Chapter 14 Fractions 2

Dividing with fractions

Key maths idea

How many halves are there in two whole oranges?

If you divide two oranges into halves, you get four halves: $2 \div \frac{1}{2} = 2 \times \frac{2}{1} = 4$

If you divide half an orange into two equal parts, you get quarters: $\frac{1}{2} \div 2 = \frac{1}{2} \div \frac{2}{1} = \frac{1}{2} \times \frac{1}{2} = \frac{1}{4}$

To divide by a fraction, you multiply by the reciprocal of the fraction you are dividing by.

We can use the same rule to divide a fraction by another fraction.

Example 1

What is $\frac{3}{4} \div \frac{1}{8}$?

$\frac{3}{4} \div \frac{1}{8} = \frac{3}{4} \times \frac{8}{1}$

$= \frac{3 \times 8}{4 \times 1}$

$= \frac{24}{4}$

$= 6$

This is really asking: How many equal groups of one eighth can we make if we have three quarters?

Three quarters

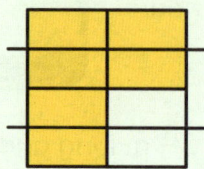
Quarters divided into halves to make eighths

1 Divide. Show at least one step in your working.

a $\frac{1}{3} \div 4$ b $\frac{1}{5} \div 6$ c $\frac{1}{9} \div 4$ d $\frac{3}{10} \div 5$

e $12 \div \frac{1}{3}$ f $9 \div \frac{1}{6}$ g $2 \div \frac{2}{3}$ h $\frac{3}{8} \div 5$

i $4 \div \frac{4}{5}$ j $\frac{4}{5} \div 4$ k $\frac{3}{8} \div 2$ l $2 \div \frac{3}{8}$

2 Divide. Show your working.

a $\frac{1}{2} \div \frac{1}{2}$ b $\frac{1}{2} \div \frac{1}{4}$ c $\frac{1}{2} \div \frac{1}{10}$ d $\frac{1}{2} \div \frac{1}{8}$

e $\frac{1}{3} \div \frac{2}{3}$ f $\frac{1}{8} \div \frac{3}{8}$ g $\frac{5}{12} \div \frac{1}{6}$ h $\frac{2}{1} \div \frac{1}{4}$

i $\frac{3}{5} \div \frac{2}{3}$ j $\frac{9}{15} \div \frac{2}{3}$ k $\frac{4}{7} \div \frac{5}{2}$ l $\frac{3}{10} \div \frac{3}{2}$

Problem solving

1 Nesha has $\frac{3}{4}$ of a pizza. She shares it with four friends. What fraction of the whole pizza does each person receive?

2 Tricia buys a 12-kg bag of cat food. Her cats eat $\frac{1}{3}$ of a kilogram of food per day. How long will the bag last her?

What did you learn?

Look back at the work you did in this chapter. Rate your progress.

1 = I cannot do this. **2** = I need more practice. **3** = I understand it and feel confident.

Can you:
- add and subtract fractions and mixed numbers?
- multiply fractions and mixed numbers?
- divide fractions by whole numbers and by other fractions?
- solve problems involving fractions and mixed numbers?

Review: Fractions 2

Key terms and concepts

1. Complete these statements in your own words to summarise what you learned in this topic.
   a. When you add or subtract fractions, it is important to …
   b. Mixed numbers are easier to work with if you …
   c. When you multiply a fraction by another fraction, you must remember to …
   d. A reciprocal is …
   e. You do not ever really have to divide fractions because …

Quick check

1. Solve:
   a. $\frac{1}{4}$ more than $\frac{1}{3}$
   b. $\frac{1}{2}$ less than $\frac{5}{8}$
   c. $1\frac{1}{2}$ times greater than $2\frac{3}{4}$

2. Calculate. Show the method you use.
   a. $\frac{3}{5} + \frac{4}{10}$
   b. $2\frac{2}{3} + \frac{3}{10}$
   c. $\frac{3}{4} - \frac{3}{8}$
   d. $2\frac{1}{8} - \frac{5}{8}$
   e. $\frac{2}{3}$ of 48
   f. $\frac{3}{4}$ of $48
   g. $\frac{2}{5}$ of $25
   h. $\frac{5}{8} \times \frac{3}{5}$
   i. $\frac{2}{3} \times \frac{9}{10}$
   j. $\frac{4}{5} \div 3$
   k. $5 \div \frac{3}{5}$
   l. $\frac{3}{5} \div \frac{1}{2}$
   m. $\frac{2}{3} \div 9$
   n. $\frac{3}{7} \div \frac{1}{2}$
   o. $\frac{5}{4} \div \frac{1}{3}$
   p. $\frac{9}{5} \div \frac{3}{4}$

3. Answer these questions about your own work in this topic.
   a. What did you find the easiest in this topic? What made it easy for you?
   b. What did you find the most challenging in this topic? What made it challenging for you?

4. Dinesh has $\frac{3}{4}$ m of rope. How many pieces of $\frac{1}{10}$ m can he cut from this?

5. Javid mixes $\frac{3}{4}$ cup of sunflower seeds and $\frac{1}{2}$ cup of sesame seeds. He adds $\frac{1}{16}$ cup to his porridge every morning. For how many days will the mixture last?

Challenge and investigate

1. One sixth of Maleek's allowance is $25. Calculate Maleek's whole allowance.

2. Ms Maraj bought 4 bunches of 12 flowers. She made these into smaller bunches, each containing $\frac{2}{3}$ of a bunch. How many smaller bunches did she make using all the flowers?

3. A bakery uses 15 cups of flour to bake a batch of bread rolls. They need $\frac{1}{4}$ of a cup of flour for one roll. How many bread rolls do they bake if they make three batches?

4. What is the area of a rectangular patch of lawn $3\frac{1}{2}$ m wide and $5\frac{3}{4}$ m long?

5. Ato launches a rocket that travels $88\frac{1}{2}$ metres in 1 second. How far will it travel in 9 seconds?

6. Amari needs $2\frac{1}{4}$ tins of paint to paint a wall. Alvin needs to paint a wall that is $1\frac{1}{2}$ times greater in area than Amari's wall. How much paint will Alvin need?

7. On Saturday, Adara spent $1\frac{3}{5}$ hours at the beach in the morning and another $1\frac{1}{2}$ hours at the beach in the afternoon. How much time did she spend at the beach on Saturday?

SECTION 9

Chapter 15 Decimals 2

In this chapter, you will:
- revisit decimals and decimal place value
- multiply with decimals
- divide decimals by whole numbers
- multiply and divide decimals by 10 and 100
- solve mixed problems involving decimals.

Starting point

1. **a** Write three questions about the decimals in the picture.
   b Give your questions to a partner to answer.

2. Besides prices like these, where else have you seen decimals being used in daily life? Make a list.

3. Look at the cut pieces of fruit in the picture.
   a Which pieces show 0.5 of a full fruit?
   b Which pieces show 0.25 of a full fruit?

Real-life maths

Athletes' finishing times at sporting events are recorded very accurately. The athletes pass through an electronic finish line – a thin beam of light – that records the time to the nearest 100th of a second. Jereem Richards holds the record in Trinidad and Tobago for the 150 m (14.75 seconds) and 200 m (20.03 seconds). His personal best time over 200 m (overseas) is 19.80 seconds.

Repeating patterns

Key maths idea

Decimals are numbers that use **place value** and a decimal point to show fractions.
The number 14.75 is composed of:
14   A whole number with 1 ten and 4 ones
7 **tenths** ($\frac{7}{10}$)
5 **hundredths** ($\frac{5}{100}$)

Key words
place value
tenths
hundredths

Repeating patterns

(continued)

We can show the number on a place value table, like this:

Tens	Ones	.	tenths	hundredths
1	4		7	5

We read 14.75 as 'fourteen point seven five'.
We can write 14.75 in expanded form, like this:
$10 + 4 + \frac{7}{10} + \frac{5}{100}$
14.75 can be written as an equivalent mixed number: $14\frac{75}{100}$
This is $14\frac{3}{4}$ in its simplest form.

1 Answer true or false for each statement.
   - **a** 0.8 means 8 of 10 equal parts
   - **b** 0.6 is the same as 0.60
   - **c** 0.3 is equivalent to $\frac{3}{10}$
   - **d** 0.20 is equivalent to 2 hundredths
   - **e** 0.25 has two decimal places
   - **f** 5 = 5.00
   - **g** $\frac{0}{65}$ is the same as $\frac{65}{100}$
   - **h** The value of 3 in 0.23 is $\frac{3}{100}$
   - **i** 0.08 is greater than 0.7
   - **j** $6\frac{5}{100}$ can be written as 6.05

2 Write these decimals as common fractions.
   - **a** 0.1
   - **b** 0.3
   - **c** 0.8
   - **d** 0.03
   - **e** 0.09
   - **f** 0.01
   - **g** 0.24
   - **h** 0.20
   - **i** 0.99
   - **j** 0.25

3 Write each decimal as a mixed number. Simplify the fraction part if possible.
   - **a** 1.5
   - **b** 3.05
   - **c** 8.25
   - **d** 3.10
   - **e** 4.26
   - **f** 1.85
   - **g** 2.75
   - **h** 3.6
   - **i** 4.35
   - **j** 2.80

4 Convert each set of fractions to decimals. Then write them in **ascending** order.
   - **a** $\frac{2}{5}, \frac{1}{4}, \frac{1}{2}, \frac{12}{50}$
   - **b** $\frac{3}{5}, \frac{12}{15}, \frac{40}{100}, \frac{6}{15}$
   - **c** $\frac{4}{5}, \frac{1}{20}, \frac{1}{2}, \frac{3}{4}$

Mental maths

The jumps on the number line show that $0.2 + 0.2 + 0.2 + 0.2 = 0.8$, so $0.2 \times 4 = 0.8$

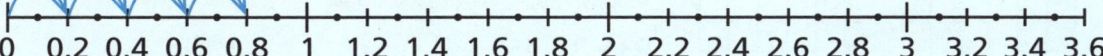

1 Use the number line to find:
   - **a** 3×0.4
   - **b** 2×0.6
   - **c** 8×0.1
   - **d** 7×0.2
   - **e** 0.3×4
   - **f** 0.5×2
   - **g** 0.3×5
   - **h** 0.7×3

2 Look at your answers. What patterns can you see?

Section 9 **Number** Chapter 15 Decimals 2

Multiplying decimals by whole numbers

Key maths idea

Multiplying with decimals is like multiplying with whole numbers, as long as you keep track of place value and insert the decimal point correctly in the answer.

Key word
commutative

Example 1
What is 0.1 × 4?

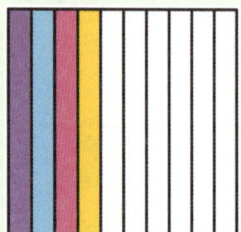

This model shows a whole (1) divided into tenths. We can colour four 0.1 sections of the square.

The model shows that the answer is less than one whole.

$0.1 = \frac{1}{10}$, so we can also work this out by converting the decimal to a proper fraction and using what we know about multiplying fractions to find the answer.

$0.1 \times 4 = \frac{1}{10} \times 4$

$= \frac{1}{10} \times \frac{4}{1} = \frac{4}{10}$

$\frac{4}{10} = 0.4$

When you multiply by a decimal, the answer will have the same number of decimal places as there are in the question.

Example 2
0.12 × 3 = ☐

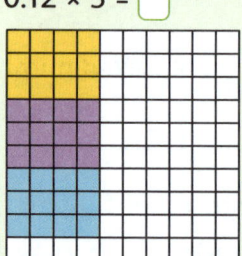

This model shows a whole (1) divided into hundredths.

Three sections of 0.12 are coloured.

Hint
Remember: 0.12 × 3 = 3 × 0.12, because multiplication is **commutative** (you can swap the order of the numbers in the calculation).

36 out of 100 blocks are coloured.

So, $0.12 \times 3 = \frac{36}{100} = 0.36$

You can also calculate this as: $\frac{12}{100} \times 3$

$\frac{12}{100} \times 3 = \frac{12}{100} \times \frac{3}{1} = \frac{36}{100} = 0.36$

Example 3
0.53 × 3 = 1.59

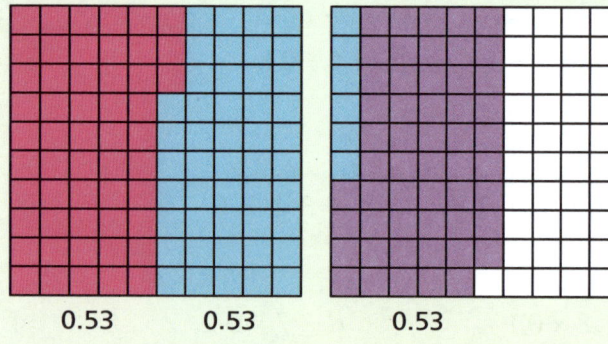
   0.53   0.53   0.53

Hint
You can also think of the answer as one whole block, five ten strips and nine ones.

53 × 3 = 159
0.53 × 3 = 1.59
2 decimal places

Talking maths

Why are place value positions important when you calculate with decimals? Share your ideas with a partner.

Multiplying a decimal by a decimal

1. Calculate. Use the method you find the easiest.
   - a 3 × 0.2
   - b 2 × 0.6
   - c 5 × 0.6
   - d 4 × 0.3
   - e 0.8 × 3
   - f 0.7 × 4
   - g 0.9 × 6
   - h 0.8 × 9
   - i 0.12 × 2
   - j 0.25 × 3
   - k 0.02 × 3
   - l 0.15 × 2

2. Calculate. Use a calculator to check your answers.
   - a 3.1 × 2
   - b 2.5 × 3
   - c 1.4 × 2
   - d 3.7 × 3
   - e 12.1 × 3
   - f 9 × 10.5
   - g 3 × 2.7
   - h 4 × 1.9
   - i 2 × 1.51
   - j 3 × 1.32
   - k 5 × 2.54
   - l 1.25 × 6

3. These calculations are incorrect. Rewrite them with the decimal point in the correct position.
   - a 12 × 08 = 96
   - b 1.5 × 3 = 45
   - c 012 × 7 = 0.84
   - d 6 × 0.19 = 114

Problem solving

1. Onika needs 7 strips of wood, each 2.4 m long. What is the total length she needs and the cost if the strips are $5 per metre?
2. Amar earns $22.85 per hour. What will he earn in 8 hours?
3. A transport company charges $45.30 per day to transport students to and from school. If there are 51 days in the school term, what is the total cost of the transport?
4. A sprinter runs 10.25 metres per second. How far will they run in 8 seconds?

Multiplying a decimal by a decimal

Key maths idea

We can use models to find a rule for multiplying a decimal by a decimal.

Each block represents $\frac{1}{100}$ or 0.01 of the whole square.

A rectangle 0.6 units long and 0.4 units wide is shaded.

The rectangle covers 24 blocks. This is $\frac{24}{100}$ or 0.24 of the whole square.

We know that area = length × width. So, 0.4 × 0.6 = 0.24 square units.

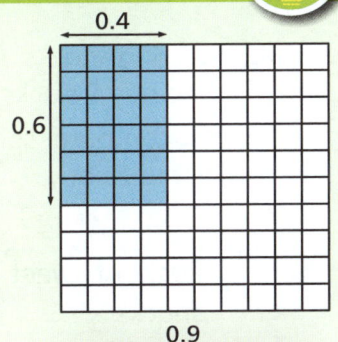

A rectangle of 0.9 × 0.7 is shaded on this model.
We can see that the area of the rectangle is $\frac{63}{100}$ or 0.63.
So, 0.9 × 0.7 must be 0.63
We can also work this out without the grid.
We know that 9 × 7 = 63
0.9 × 0.7 = 0.63 ← 2 decimal places
1 decimal place each
We can also check this using equivalent common fractions: $\frac{9}{10} \times \frac{7}{10} = \frac{63}{100}$ = 0.63

When you multiply decimals, the number of decimal places in the answer must be the same as the total number of decimal places in the question. This rule allows us to multiply as if there were no decimal points. We then use the rule to place the decimal point in the correct position in the answer.

Section 9 Number Chapter 15 Decimals 2

(continued)

Example 1

$0.7 \times 0.5 = \square$ $\qquad$ $7 \times 5 = 35$

There are two decimal places in the question, so there must be two decimal places in the answer: $0.7 \times 0.5 = 0.35$

You can test this idea using your calculator.

Example 2

What is the area of a rectangular floor 2.4 m wide and 5.6 m long?

Work out 24×56

×	20	4
50	1000	200
6	120	24

= 1200
= 144
= 1344

Use the rule to insert the decimal point in the answer:

$2.4 \times 5.6 = 13.44$ ← 2 decimal places

1 decimal place each

1 Do each multiplication. Use a calculator to check that you have the decimal point in the correct position in your answers.

- **a** 0.4×0.4
- **b** 0.2×0.6
- **c** 0.7×0.1
- **d** 0.5×0.9
- **e** 0.3×0.8
- **f** 0.7×0.8
- **g** 0.2×0.2
- **h** 0.6×0.5
- **i** 0.3×0.9
- **j** 0.6×0.6
- **k** 0.9×0.1
- **l** 0.9×0.7

2 Calculate. Show your working.

- **a** 3.1×9.2
- **b** 4.5×4.5
- **c** 12.2×1.8
- **d** 9.5×1.2
- **e** 6.7×0.9
- **f** 4.2×1.9
- **g** 10.1×0.9
- **h** 12.3×1.8

Full STEAM ahead Using spreadsheets

Jemila is buying snacks. A pack of sweet snacks is $20.50 and a pack of savoury snacks is $10.99. She plans to buy 5 packs and spend no more than $80.

1 Make a computer spreadsheet like the one below to show what combinations of snacks Jemila can afford to buy.

	A	B	C	D	E
1	Packs of sweet snacks	Packs of savoury snacks	Total cost of sweet snacks	Total cost of savoury snacks	Total cost of 5 packs
2	0	5			
3	1	4			
4	2	3			
5	3	2			
6	4	1			
7	5	0			

2 What rules (algorithms) can you use to calculate the totals in columns C, D and E?

3 Use the programme (or a calculator) to complete columns C, D and E.

4 What combinations are possible for Jemila to buy? List them.

5 Jemila gets another $50, so she decides to buy 8 packs and spend no more than $120. Make a new spreadsheet to show her options.

Dividing a decimal by a whole number

Key maths idea

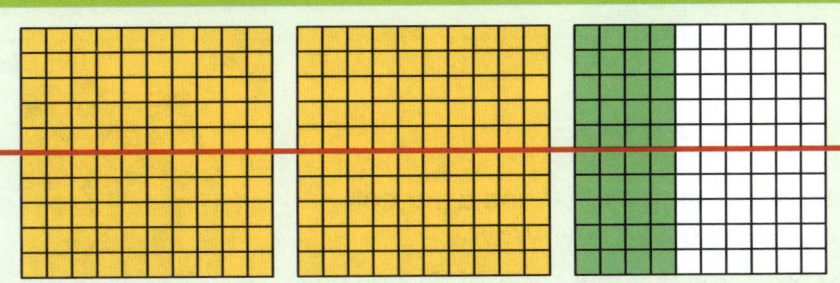

The diagram shows 2.4. Each small block represents 0.01 or $\frac{1}{100}$ of the whole.

The line divides 2.4 into two equal parts. You can see that: 2.4 ÷ 2 = 1.2

Count squares on the diagram to check these division facts.

2.4 ÷ 3 = 0.8	2.4 ÷ 4 = 0.6	2.4 ÷ 5 = 0.48	2.4 ÷ 6 = 0.4
2.4 ÷ 8 = 0.3	2.4 ÷ 10 = 0.24	2.4 ÷ 12 = 0.2	2.4 ÷ 24 = 0.1

You can see that some divisions result in one decimal place in the answer, but others result in two decimal places.

To keep track of the decimal point when you divide a decimal by a whole number, write the decimal point in the answer directly above the decimal point in the question. Then divide as you would for whole numbers.

Example 1

126.8 ÷ 2

```
      63.4
    _____
  2 | 126.8
```

Example 2

142.8 ÷ 12

```
       11.9
     _____
  12 | 142.8
     - 12 ↓
     _____
        22
      - 12 ↓
      _____
         108
       - 108
       _____
```

1 Divide. Try to work out the answers mentally.

a 0.3 ÷ 1   b 0.8 ÷ 2   c 0.9 ÷ 3   d 0.10 ÷ 2
e 1.2 ÷ 4   f 3.6 ÷ 9   g 4.2 ÷ 7   h 3.5 ÷ 7

2 Divide. Show any working that you do.

a 9.6 ÷ 2   b 5.6 ÷ 8   c 14.1 ÷ 2   d 11.6 ÷ 8
e 5.4 ÷ 6   f 9.27 ÷ 3   g 30.25 ÷ 5   h 82.8 ÷ 9

3 Calculate the following amounts of money. Use a calculator to check your answers.

a $13.50 ÷ 6   b $124.80 ÷ 4   c $24.99 ÷ 3   d $132.74 ÷ 2
e $12.64 ÷ 8   f $18.75 ÷ 15   g $455.80 ÷ 4   h $234.56 ÷ 32

4 The perimeter of each regular polygon is given. Work out the length of one side.

a 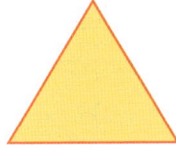   b    c   d

Perimeter: 18.75 cm   Perimeter: 14.44 m   Perimeter: 6.5 cm   Perimeter: 5.7 m

Section 9 Number Chapter 15 Decimals 2

Talking maths

1. How is dividing a decimal by a whole number similar to dividing a whole number by a whole number? Give examples to support your answer.

Problem solving

1. **a** A store sells a pack of 6 bottles of water for $7.50 and a pack of 8 bottles for $9.60. Which pack is the better buy? Why?

   b A case of 24 bottles costs $27.60. If you buy four cases of 24, you pay $105.60. How much do you save per bottle of water if you buy four cases?

2. A tourist bought three T-shirts at a market for $36.20 and got a fourth T-shirt free. What did they end up paying per T-shirt?

3. A 50-ml bottle of suntan lotion costs $41.50 and a 60-ml bottle costs $48.60. Which is the most economical to buy? Why?

4. Mario puts 25 litres of gas in his car. He pays $193.75. How much does 1 litre of gas cost?

Multiplying and dividing by 10 and 100

Key maths idea

When you multiply any number by 10, you make it one place greater. When you multiply by 100, you make the number two places greater. For example: 4 × 10 = 40 and 4 × 100 = 400.

The place value table shows how this rule applies to decimals.

	Hundreds	Tens	Ones	.	tenths	hundredths
			1	.	2	5
× 10		1	2	.	5	

When you multiply a decimal by 10, the digits all move one place to the left: 1.25 × 10 = 12.5

	Hundreds	Tens	Ones	.	tenths	hundredths
			1	.	2	5
× 100	1	2	5	.		

When you multiply a decimal by 100, the digits all move two places to the left: 1.25 × 100 = 125

Division is the inverse of multiplication. When you divide any number by 10, you make it one place smaller. When you divide by 100, you make the number two places smaller.

	Hundreds	Tens	Ones	.	tenths	hundredths
	1	6	9	.		
÷ 10		1	6	.	9	

When you divide a decimal by 10, the digits all move one place to the right: 169 ÷ 10 = 16.9

	Hundreds	Tens	Ones	.	tenths	hundredths
	1	6	9			
÷ 100			1	.	6	9

When you divide a decimal by 100, the digits all move two places to the right: 169 ÷ 100 = 1.69

Multiplying and dividing by 10 and 100

1. Find the incorrect number sentence in each row. Write the correction.

   a. 7.54 × 100 = 7540   7.54 × 10 = 75.4   75.4 × 10 = 754
   b. 7500 ÷ 10 = 750   7500 ÷ 100 = 750   75 ÷ 10 = 7.5
   c. 19.2 × 10 = 192   13.6 × 100 = 136   12.7 × 10 = 127
   d. 34 ÷ 10 = 3.4   45 ÷ 10 = 4.5   0.9 ÷ 10 = 9
   e. 23 × 10 = 230   2.3 × 10 = 23   2.3 × 100 = 2300

Hint

Remember: a number machine takes a number (the input) and performs an operation on it to get another number (the output).

2. Work out the missing numbers in each number machine.

   a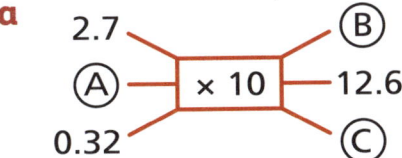
   b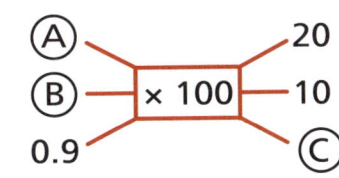
   c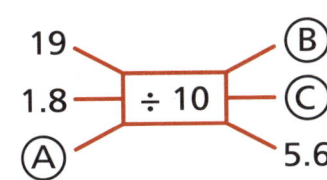

3. The operation is missing from each number machine. Work out what it should be and write the missing numbers.

   a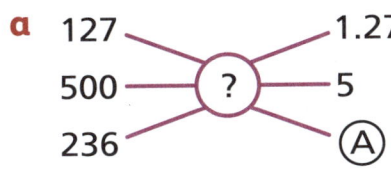
   b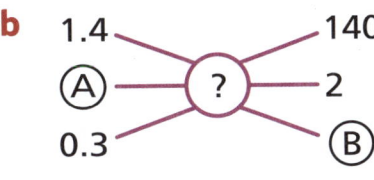
   c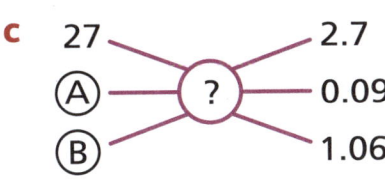

Mental maths

Do these calculations mentally. Use a calculator to check your answers.

   a 0.5 × 10   b 0.5 × 100   c 3.1 × 10   d 3.1 × 100
   e 4.2 × 10   f 4.2 × 100   g 20.4 × 10   h 20.4 × 100
   i 121 ÷ 10   j 121 ÷ 100   k 2009 ÷ 10   l 2009 ÷ 100
   m 25.3 ÷ 10   n 29.3 ÷ 10   o 14.9 ÷ 10   p 0.8 ÷ 10

What did you learn?

Look back at the work you did in this chapter. Rate your progress.
1 = I cannot do this. 2 = I need more practice. 3 = I understand it and feel confident.

Can you:
- multiply decimals by whole numbers?
- multiply decimals by other decimals?
- divide decimals by whole numbers?
- multiply and divide decimals by 10 and 100?
- solve real-life problems involving decimals?

Section 9 Number Chapter 15 Decimals 2

Review: Decimals 2

Key terms and concepts

1 Give examples using numbers and diagrams to show that you understand each of these concepts and to summarise what you learned in this chapter.
   a Writing decimals as common fractions
   b Multiplying a decimal by a whole number
   c Multiplying a decimal by another decimal
   d Dividing a decimal by a whole number
   e Multiplying and dividing decimals by 10 and 100

Quick check

1 Multiply:
   a 4.5 × 10
   b 2.4 × 8
   c 3.12 × 2
   d 1.25 × 6
   e 1.2 × 0.9
   f 2.1 × 0.2
   g 0.8 × 0.8
   h 2.8 × 2.5

2 Divide:
   a 0.9 ÷ 3
   b 0.8 ÷ 2
   c 1.2 ÷ 6
   d 0.24 ÷ 3
   e 1.25 ÷ 5
   f 0.72 ÷ 8
   g 33.99 ÷ 3
   h 135.54 ÷ 9

3 Work out the missing operations in each set. Write the letter and the operation.

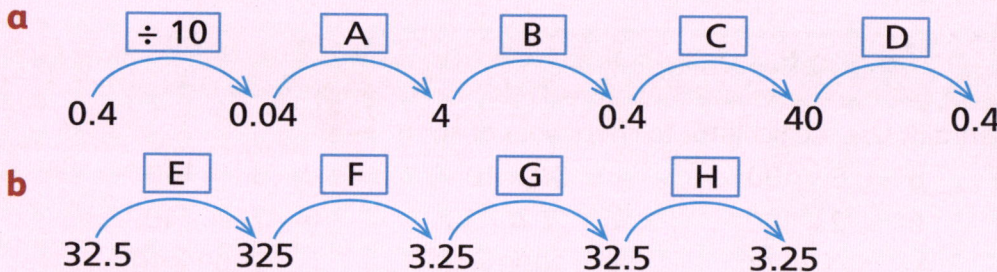

Challenge and investigate

1 It takes the Earth 364.25 days to complete one revolution around the Sun. How many days does it take for the Earth to complete three revolutions?

2 Mona cuts a 4.75 m long wire into five equal parts. How long is each part?

3 One calculation is incorrect. Find it and do the correction.

$$5\overline{)31.5}^{\,6.3} \qquad 5\overline{)315}^{\,63} \qquad 5\overline{)3.15}^{\,6.3}$$

4 The average mass of these four bags of food is 2.15 kg.

The average mass of bag A and B is 2.2 kg.
Bag D is 3 times heavier than bag C.
What is the mass of bag C and bag D?
Convert each mass to grams.

A   B    C    D

SECTION 10

Chapter 16 Solids and plane shapes 2

In this chapter, you will:
- revise what you learned about solids and plane shapes in Standard 3
- describe and classify solids and plane shapes according to their properties (sides, angles and vertices)
- identify, classify and sort quadrilaterals according to their properties
- learn about the cross-sections of different solids
- identify lines of symmetry and solve problems involving symmetry.

Starting point

1. Work with a partner. Find the following shapes on the photograph of electricity pylons.
   - a parallel lines
   - b a pentagon
   - c a right-angled triangle
   - d a scalene triangle
   - e a four-sided polygon that is not a square or rectangle

A pylon is a tall structure designed to support power lines

Explore plane shapes

Key maths idea

Plane shapes are **2-D** because they have only two dimensions: length and width. When you describe the properties of a 2-D shape, you should identify:
- how many **sides** and **vertices** it has
- the kinds of **angles** formed at its vertices
- whether any of the sides are equal in length
- how many right angles it has.

Polygons have straight sides. In a **regular** polygon, the sides are all the same length and the sizes of the angles at the vertices are equal.

Key words
plane shape
2-D
side
vertex / vertices
angle
polygon
regular

  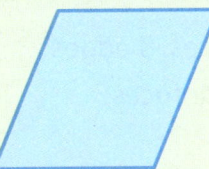 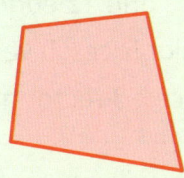

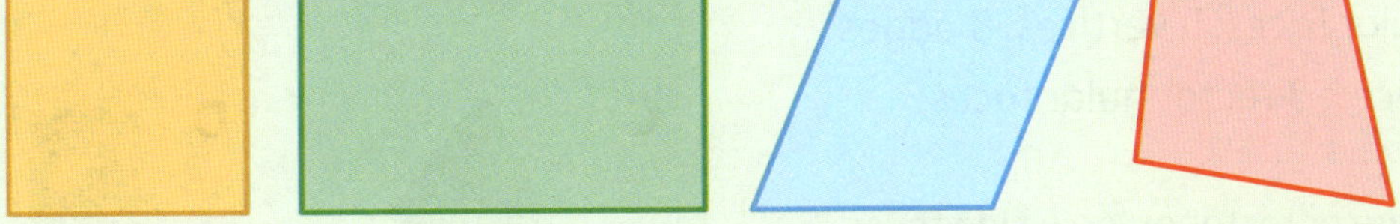

All of these shapes have the same number of sides and vertices, but their angles are different. Only one is a regular polygon. Can you see which one it is?

169

Section 10 Geometry Chapter 16 Solids and plane shapes 2

> **Talking maths**
>
> 1. Look at these words: trio, tricycle, trilogy, triathlon. Discuss with a partner what they mean, and what they tell you about the prefix tri-.
> 2. Why is it impossible to create a polygon with only two sides?
> 3. In a group of three, use pencils or pens to form the different kinds of triangles you learned about in Chapter 5. If you cannot remember all the kinds of triangles, turn back to page 49.
> 4. What properties does a square have that not all rectangles have?

1. Write the letter of the shape that does not belong in each set. Identify two properties it has that are different from the others in the set.

   a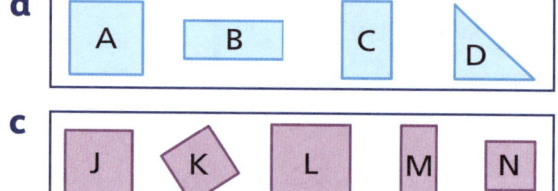

   b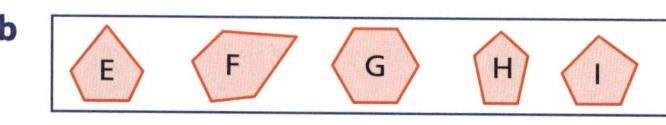

   c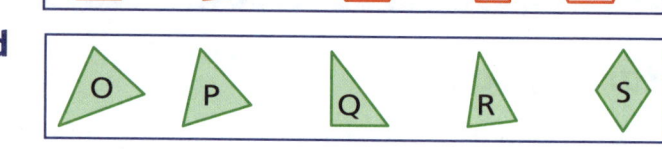

   a   b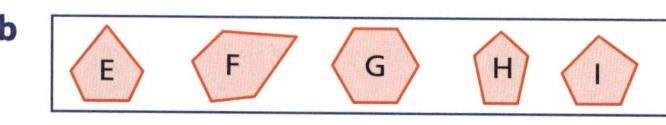

   c (purple shapes J K L M N) d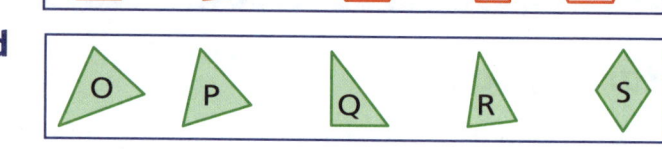

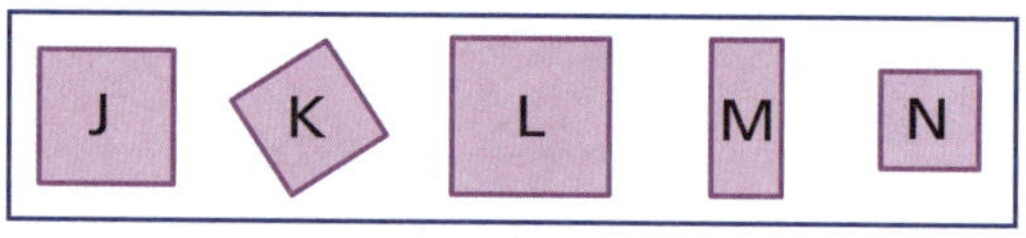

2. Draw these plane shapes.
   a A square with each side 6 cm long
   b A rectangle with width 2.5 cm and length 3 cm
   c A scalene triangle with its longest side 8 cm
   d A right-angled isosceles triangle

Explore solids

> **Key maths idea**
>
> A **solid** has three dimensions: length, width and height. Spheres, cones, cubes and pyramids are all examples of solids. We classify solids according to their properties:
> - whether they have curved **surfaces**, flat **faces** or both
> - the number of **edges**, faces and vertices
> - the shape of the base and other faces
> - the angles of the faces.
>
> **Key words**
> solid
> surface
> face
> edge

1. Write the letter of the diagram and the name of the solid that matches each description.

   a 6 square faces, 8 vertices and 12 edges
   b 1 square face, 4 triangular faces, 5 vertices, 8 edges
   c 2 identical triangular faces, 3 rectangular faces, 6 vertices, 9 edges
   d 2 end faces that are square or rectangular, 4 side faces that are rectangular, 8 vertices and 12 edges
   e 1 circular face and a curved surface that ends at a point
   f 2 circular end faces joined by a curved surface, no vertices

A    B

C    D

E    F

Prisms and pyramids

Key maths idea

A **prism** is a solid that has two parallel end faces. The end faces are joined by rectangular faces. We name prisms according to the shape of the end faces. The end face is also called the **base**.

Key words
prism
base
pyramid
apex

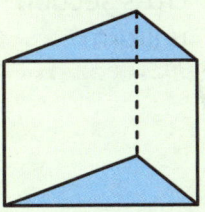
triangular prism

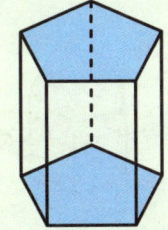
pentagonal prism

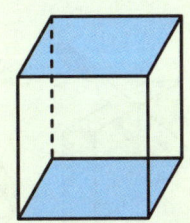
rectangular prism

A **pyramid** is named according to the shape of its base. Pyramids have a polygon as a base, and triangular side faces that meet at a point called the **apex**.

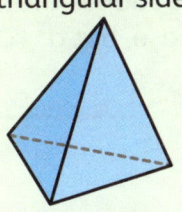
triangular pyramid

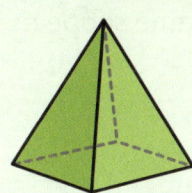
square-based pyramid

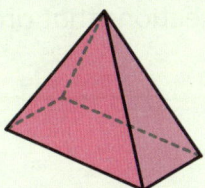
rectangular-based pyramid

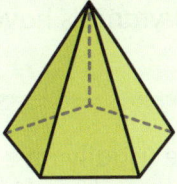
pentagonal pyramid

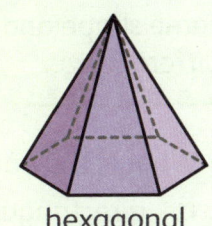
hexagonal pyramid

1. Write the name of the prism that has:
   a. three rectangular faces and two triangular faces
   b. five rectangular faces and two pentagonal faces.

2. What is another name for a rectangular prism?

3. Omar says that only rectangular prisms have right angles on their faces. Gary says this is a mistake, and that all prisms have faces with right angles.
   a. Who is correct, and why?
   b. What kinds of pyramids have right angles?

4. If a rectangular prism has all its edges the same length, what is it called?

5. a. What is the relationship between the base of a pyramid and the number of triangular side faces it has?
   b. Is this the same or different for prisms? Why?

6. How many pairs of parallel lines do you see on:
   a. a triangular prism?
   b. a pentagonal prism?
   c. a rectangular prism?

7. What plane shape always occurs as more than one of the faces of:
   a. a prism?
   b. a pyramid?

Hint
Think about the solids you have already learned about.

Section 10 Geometry Chapter 16 Solids and plane shapes 2

Cross-sections

> **Key maths idea**
>
> Imagine that you could slice through a solid to make two smaller shapes. This kind of cut splits the solid into smaller solids, and it also forms another face on each of the smaller solids. The new face formed by the cut is called the **cross-section**.
>
> **Key words**
> cross-section
> uniform
>
>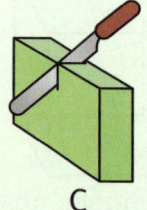
>
>   A   B   C   D   E   F
>
> If you slice a prism or a pyramid parallel to its base, you make a cross-section that is the same shape as the base. Prisms have **uniform** cross-sections. That means that every cross-section is the same shape and size. Pyramids have cross-sections that are the same shape as their base, but a different size.

1. What is the name we usually give to:
   a. a rectangular-based prism?
   b. a square-based prism with all other faces square?

2. Name three shapes that have uniform cross-sections.

3. Copy and complete this table for the shapes above.

Solid	A	B	C	D	E	F
Name						
Number of vertices						
Number of edges						
Number of faces						
Shape of base						
Shape of cross-section parallel to base						

> **Talking maths**
>
> What do you notice about the cross-sections of cylinders, cones and spheres?
>
> What similarities and differences do you notice about these, compared to the prisms and pyramids on the previous page?
>
> **Hint**
> You will need to identify the cross-sections of solids. That means you need to recognise and name the shape of the face formed by the slice.
>
>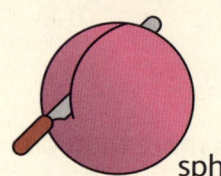
>
>   cylinder   cone   sphere

Investigating solids and plane shapes

Full STEAM ahead   Making shapes

You will need:
- pipe cleaners
- wire
- toothpicks
- pallet sticks or cotton buds (to form the edges)
- tape or adhesive putty (to join the edges)
- beads (to mark the vertices).

1   Work in groups. Collect your materials and make:
   a   two cubes of different sizes
   b   two different cuboids
   c   a cylinder
   d   a square-based pyramid
   e   a triangular-based pyramid
   f   a triangular prism.

Problem solving

1   What shape did each student make?
   a   Nikkita cut out two circles the same size. She rolled a rectangular strip into a tube and used the circles as parallel end faces.
   b   Billy cut out four identical triangles and used them to construct a solid.
   c   Lara cut out six squares and used them to construct a solid shape.

2   In each picture, you can see two different cuts: cut A and cut B. Imagine that the knife will complete the cut all the way through to the other side. For each cut:
   - say whether it is parallel to the base of the solid or its height
   - draw the shape of the cross-section it will produce.

   a   b

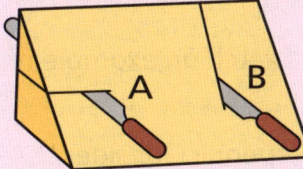   

Hint
When you cut along a line parallel to the height of a solid, you get a different cross-section from when you cut parallel to the base.

173

Section 10 Geometry   Chapter 16 Solids and plane shapes 2

Quadrilaterals

> **Key maths idea**
>
> A **quadrilateral** is a polygon with four sides and four angles.
> In a sketch of a quadrilateral, a square corner mark indicates a right angle.
> This also means that the lines meeting at this angle are **perpendicular**.
> We use short lines to mark which sides have equal lengths.
>
> **Key words**
> quadrilateral
> perpendicular
> square
> parallelogram
> rhombus
> rectangle
> kite
> trapezium
>
>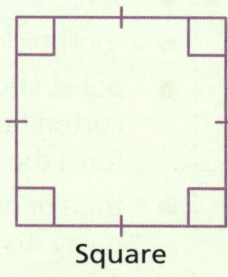
> Square
>
>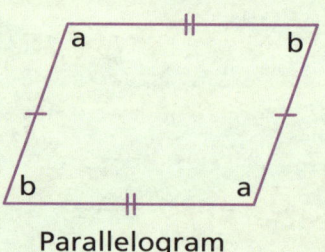
> Parallelogram
>
> A **square** is the only regular quadrilateral. Its sides are all equal in length. Its angles are all equal to 90°.
>
> A **parallelogram** has its opposite sides parallel and equal in length. The opposite angles are equal. We use matching letters to show equal angles.
>
>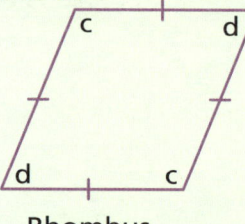
> Rhombus
>
>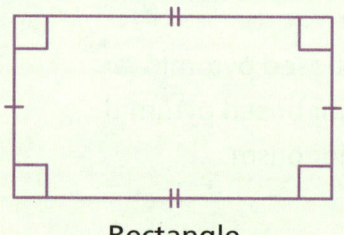
> Rectangle
>
> A **rhombus** is a parallelogram with four equal sides.
>
> A **rectangle** is a parallelogram that has a right angle at each vertex.
>
>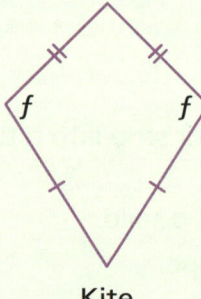
> Kite
>
>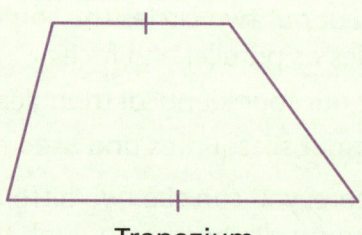
> Trapezium
>
> A **kite** has two pairs of equal sides, but they are next to each other. It has one pair of angles that are equal, where the sides of different lengths meet.
>
> A **trapezium** is a quadrilateral with one pair of opposite sides parallel.

1   a   Name the three types of quadrilaterals that are parallelograms. Sketch an example of each.
    b   Which quadrilaterals must always have perpendicular sides?
    c   Which quadrilateral is it possible to draw with only one pair of perpendicular sides?
2   What two properties do a rhombus and a square have in common?
3   What do you call a quadrilateral that has only one pair of parallel lines? Sketch an example.
4   Which two quadrilaterals have four right angles?

Quadrilaterals

5   Which of the following is **not** a name for the figure shown to the right: polygon, parallelogram, quadrilateral, trapezium.

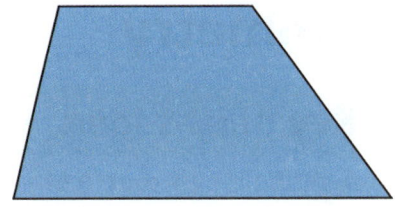

6   Mariah's teacher asked her to draw a quadrilateral with all sides equal. Which of the following could she draw: rhombus, kite, square, parallelogram, rectangle, trapezium?

7   Is it possible to draw a kite with two right angles? Fold a piece of paper to help you draw a diagram to show your answer.

8   In each set of shapes below, identify which one does not belong. Give a reason, using what you have learned about quadrilaterals.

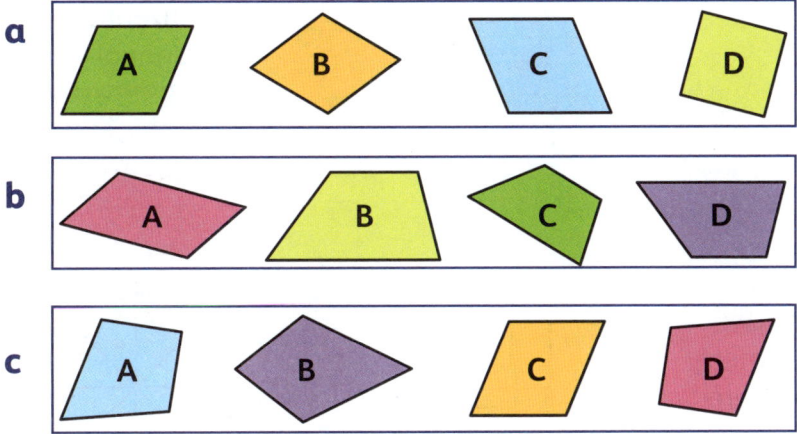

9   Identify the type of quadrilateral in each riddle.

a   Two of my angles are greater than 90°, and two of my angles are smaller than 90°, but my four sides are all equal in length.

b   I have only one right angle, and one pair of opposite parallel sides.

c   I have two pairs of sides that are equal in length, but none of them are parallel. I have one pair of opposite angles that are equal.

Problem solving

1   Abigail says that working out the perimeter of a rhombus is the same as working out the perimeter of a square. Explain why this works.

2   Jabari puts two identical triangles together to make a square. Must the triangles be isosceles, scalene or equilateral? Draw a sketch to help explain your answer.

3   A kite has a short side of 3 cm and a long side of 5 cm. Work out the perimeter.

Hint
Remember the properties of kites. Go back to Chapter 10 if you need to refresh your memory about perimeter.

Section 10 Geometry Chapter 16 Solids and plane shapes 2

Symmetry

> ### Key maths idea
>
> A shape is **symmetrical** – or it has **line symmetry** – if you can draw or imagine a line down the middle that divides it into two parts that fit exactly onto each other. This line is called a **mirror line** or line of symmetry, because the two sides mirror each other. The half shape on the other side of the mirror line is called the mirror image.
>
> **Key words**
> symmetrical
> line symmetry
> mirror line
> vertical
> horizontal
> diagonal
>
> The red lines on these shapes each show a line of symmetry:
>
>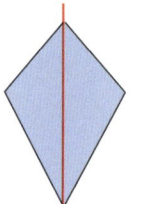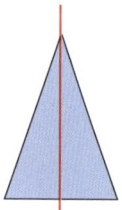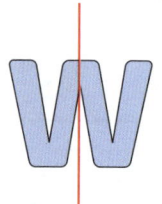
>
> A shape may have more than one line of symmetry. For example, the letter H has two lines of symmetry: a **vertical** line of symmetry and a **horizontal** line of symmetry. A square has four lines of symmetry: a vertical line, a horizontal line and two **diagonal** lines.
>
>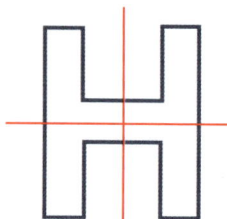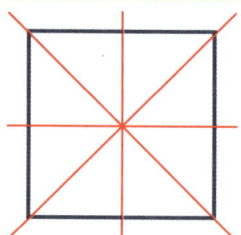

1. Look at the plane shapes below. Write the letters of the shapes that have no lines of symmetry.

   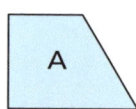 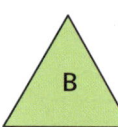  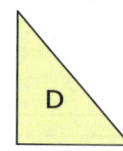 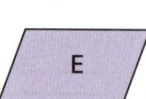
    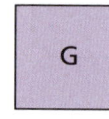 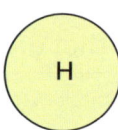  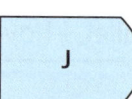

2. **a** Trace three of the shapes from question **1** that do have line symmetry. Draw the lines of symmetry.

   b One of the shapes from question **1** has so many lines of symmetry that you cannot draw them all. Identify which one it is, and explain why.

3. Half of six symmetrical shapes are given below. The red lines show the mirror lines. Trace and complete each shape.

   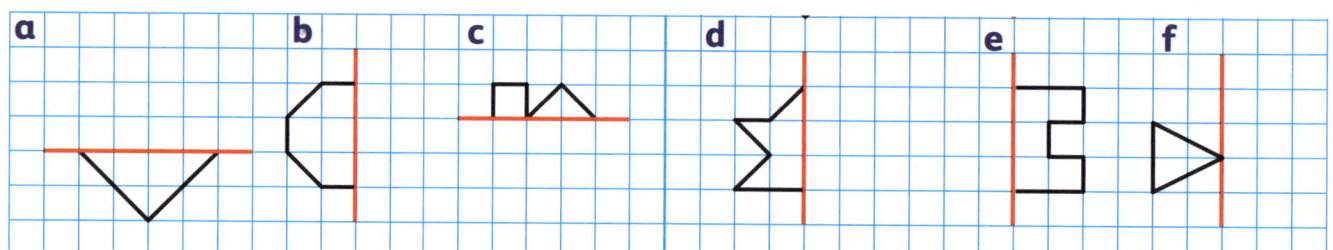

Symmetry

4 Pick the correct mirror image for each shape.

a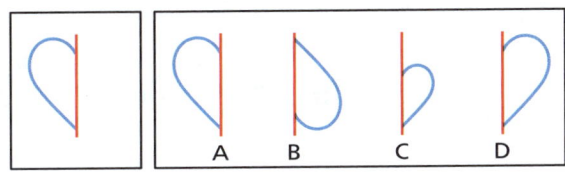

b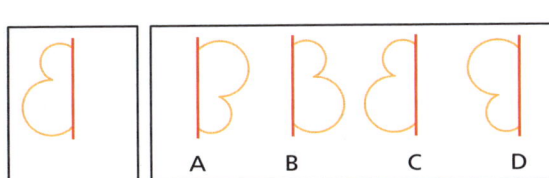

c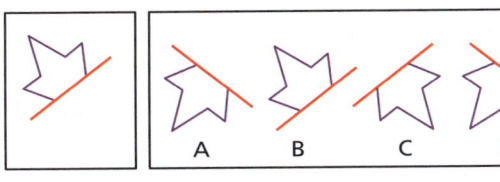

d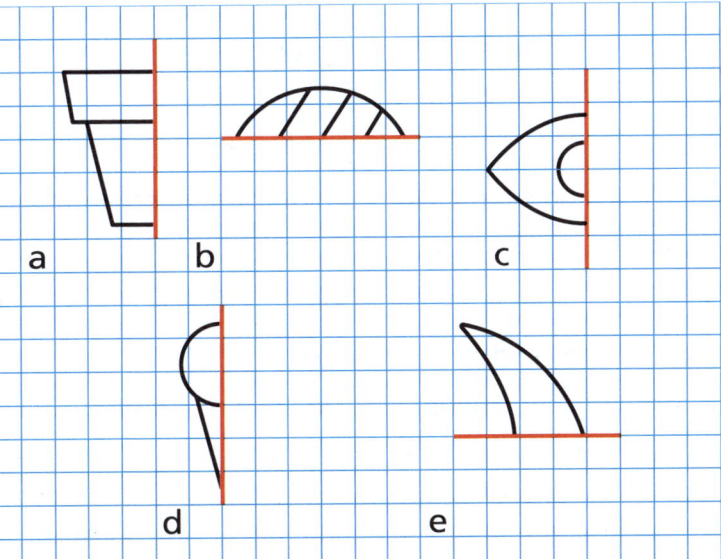

5 How many lines of symmetry do you think each shape has? Think about it and write down your guess. Then sketch the shape and mark the lines of symmetry.

- **a** rhombus
- **b** parallelogram
- **c** rectangle
- **d** scalene triangle
- **e** isosceles triangle
- **f** equilateral triangle
- **g** kite
- **h** trapezium

6 a Write your name in capital letters. Which letters have lines of symmetry?

b Write other letters of the alphabet in capitals, or look at a poster of capital letters. Identify the letters that have line symmetry.

7 The red line is the line of symmetry, but the mirror image is missing from each shape. What real-life object or shape would it make when the mirror image is complete?

Hint
Copy and complete each shape.

What did you learn?

Look back at the work you did in this chapter. Rate your progress.
1 = I cannot do this. **2** = I need more practice. **3** = I understand it and feel confident.

Can you:
- identify and classify plane shapes according to their properties?
- identify and classify solids according to their properties?
- describe cross-sections of different solids, and name the solids that have uniform cross-sections?
- construct models of solids?
- identify different kinds of quadrilaterals and describe their properties?
- solve problems involving line symmetry?

177

Section 10 Geometry Chapter 16 Solids and plane shapes 2

Review: Solids and plane shapes 2

Key terms and concepts

1. **a** Plane shapes have two dimensions: ____ and ____. A plane shape with straight sides is called a ____. The sides join at points called ____.

   b Solids have three dimensions: ____, ____ and ____. Flat faces of a solid meet at a line called an ____. The point where three or more flat faces join is called a ____.

   c A quadrilateral has four ____ and four ____. If both pairs of sides are parallel, it is a ____. If all four angles are right angles, it could be a ____ or a ____. If it has all four sides equal in length but angles unequal, it must be a ____.

2. **a** A cross-section is the face created when you slice across a **polygon / parallelogram / solid**.

   b If you slice a prism parallel to its base, you get cross-sections that are **equilateral / uniform / perpendicular**.

3. A line of ____ divides a shape into two identical shapes that can fit onto each other.

Quick check

1. Name each solid and describe the number and shape of its faces.

2. Which of the solids above would not have uniform cross-sections if sliced parallel to its base?

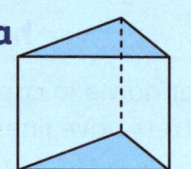
a
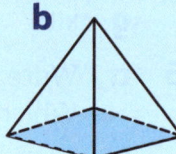
b

3. What is the main difference between a trapezium and a parallelogram?

4. Sketch an example of each of these quadrilaterals:
   a a rhombus   **b** a parallelogram   **c** a kite

5. Which of your sketches from question **4** has a line of symmetry? Add it to your sketch.

Challenge and investigate

1. Identify the quadrilateral in each description. Draw a sketch to show your answer.

   a A quadrilateral with two pairs of equal sides; each pair of equal sides shares a vertex. One pair of opposite angles are equal, and there are no parallel sides or right angles.

   b A quadrilateral with one pair of parallel lines, one right angle and no equal angles or sides of equal length.

2. Which of these solids would have uniform cross-sections if sliced parallel to their base?

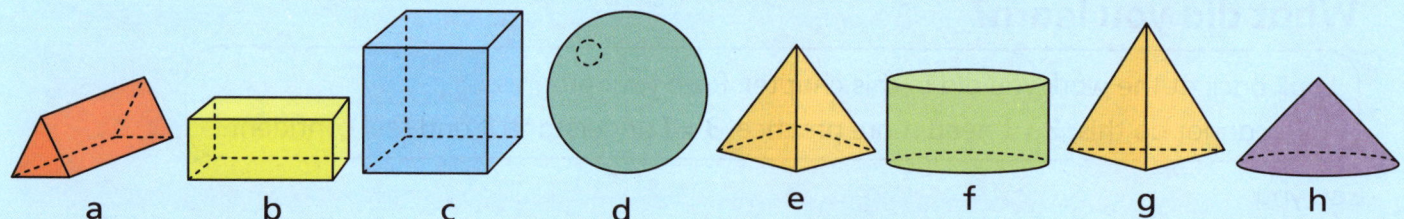

a   b   c   d   e   f   g   h

3. Draw the shape that would be formed by each of these cross-sections.

   a 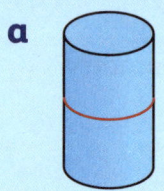   b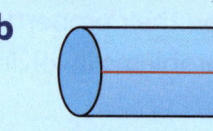

4. Sketch three capital letters that have line symmetry. Show the lines of symmetry on your sketch.

SECTION 11

Chapter 17 Perimeter and area 2

In this chapter, you will:
- use formulae to calculate the perimeter of squares, rectangles and composite shapes
- use formulae to calculate the area of squares, rectangles and composite shapes
- draw shapes given their perimeter or area
- solve real-life problems involving perimeter and area.

Starting point

Solar panels use sunlight to generate electricity. This picture shows eight arrays of panels. Each array is made up of eight panels.

1. Each array of eight panels has a thin aluminium strip around it.
   a. How can you work out the total length of the aluminium strip without measuring all the sides?
   b. Each panel is 1.2 m long and 0.5 m wide. What is the perimeter of each one?
2. a. What is the mathematical name for the amount of space that each panel covers in the array?
   b. What would you need to measure to work out the area of each array?

Talking maths

A **linear** measurement is a one-dimensional distance or length.
1. Explain why perimeter is a linear measurement but area is not.
2. How do the units for linear measurements differ from the units we use for area?

Key words
linear
formula

Perimeter

Key maths idea

Perimeter is the distance around a plane shape. We use millimetres (mm), centimetres (cm), metres (m) or kilometres (km) to measure perimeter.

You can use a rule called a **formula** to calculate the perimeter of squares and rectangles.

A square has four sides the same length, so the perimeter is the length of one side multiplied by four.

Perimeter = 4 × length of one side

Section 11 Measurement Chapter 17 Perimeter and area 2

> **(continued)**
>
> We can write this rule as a formula using letters.
> If s stands for the length of one side, then the perimeter is $4 \times s$.
> $P = 4 \times s$
> A rectangle has two pairs of sides that are equal in length.
> We can use this information to write a formula.
> Perimeter = 2 × length + 2 × width
> If we call the length l and the width w, the perimeter is $2 \times l + 2 \times w$.
> $P = 2 \times l + 2 \times w$
>
> **Hint**
> Read through Chapter 11 if you need a reminder of any of these ideas.

1 Use the appropriate formula to calculate the perimeter of each shape.

a 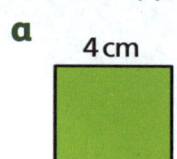 b 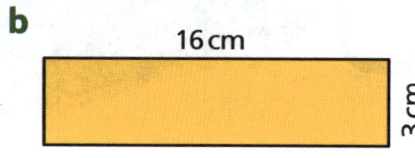 c 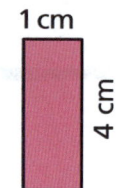 d 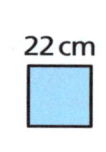 e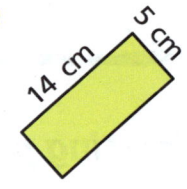

2 Which shape has the greater perimeter in each pair? Show how you decide.

a 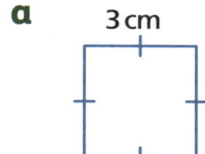 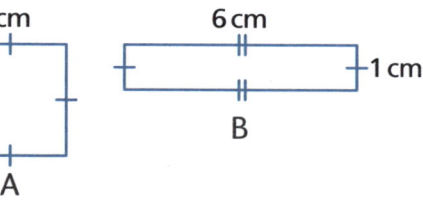   b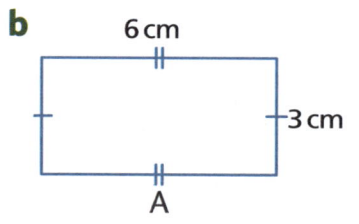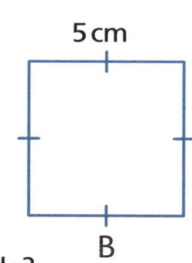

3 Which of these formulae could be for calculating the perimeter of a rectangle? Explain why the other one could not be.

$P = 4 \times l + 4 \times w$ $P = 2 \times (l + w)$

Problem solving

1 The perimeter of a rectangle is 28 cm. One side is 8 cm long. How long are the other sides?

2 Ashanti jogs around a square park every day. One side of the park is 125 m long. If she jogs around the park four times, how far will she run? Give your answer in kilometres.

3 Nigel has a square cloth with a perimeter of 6 m. What is the length of one side of the cloth?

4 The diagram shows the plan for a rectangular pool with a 1 m-wide path around it.

   a Estimate and then calculate the perimeter around the outside of the path.

   b Draw two different rectangular pools with the same perimeter as this one.

   c Does changing the shape of the pool change the perimeter of the path around it? Explain your answer.

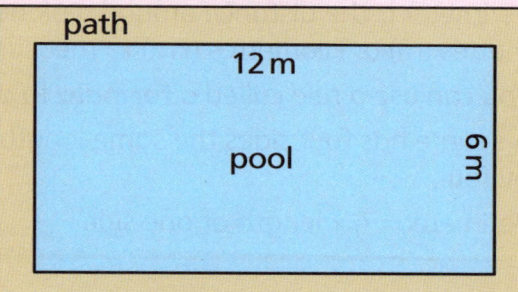

Perimeter of composite shapes

Key maths idea

To calculate the perimeter of a **composite shape**, you find the sum of all the sides. You may need to work out missing lengths before you add.

Find the missing lengths first.

We write them on the diagram and then find the sum of the lengths.

Key words
composite shape

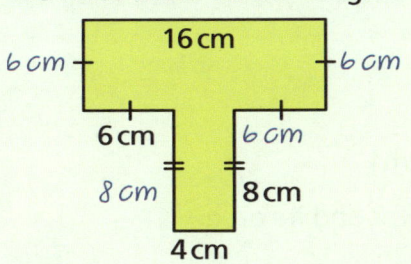

16 cm + 6 cm + 6 cm + 8 cm + 4 cm + 8 cm + 6 cm + 6 cm = 60 cm

1 Draw a rough copy of each shape. Fill in all the missing lengths and calculate the perimeters.

a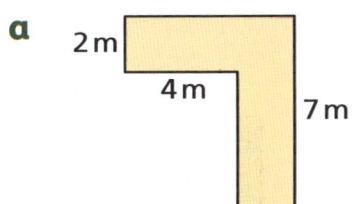

b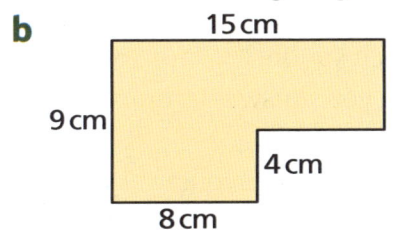

c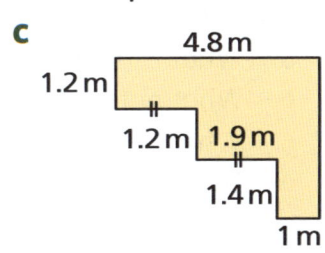

Area of squares and rectangles

Key maths idea

Area is the amount of space a shape covers. We measure area in **square units**.

One square centimetre (1 cm²) is a square 1 cm long and 1 cm wide.

You have already found the area of shapes by counting squares, and you have seen that if you know the number of units in the length and width of the shape, you can multiply to find the area without counting squares.

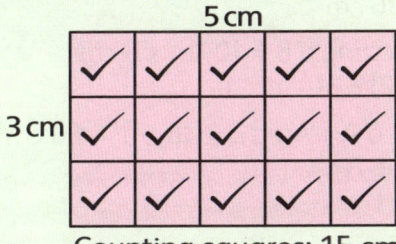

Counting squares: 15 cm²

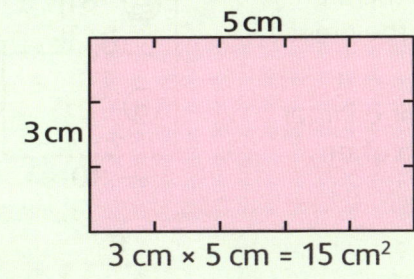

3 cm × 5 cm = 15 cm²

Key words
square units

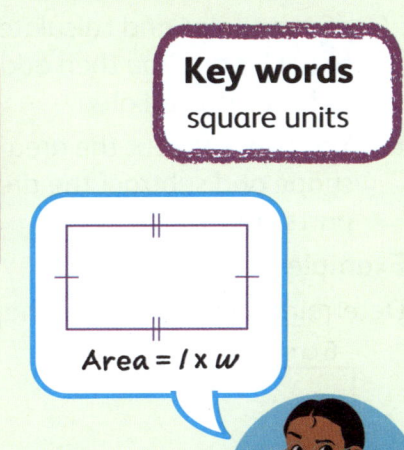

Area = $l \times w$

This leads us to the formula for finding the area of a rectangle.

Area = length × width

If we use l for length and w for width: $A = l \times w$

For a square, the length and width are equal.

Area = side × side

$A = s \times s$

Section 11 Measurement Chapter 17 Perimeter and area 2

1. Use a formula to calculate the area of each shape.

   a 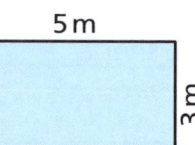   b 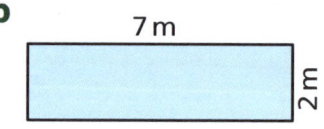   c 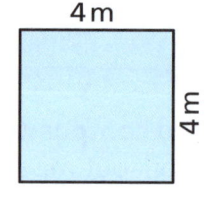   d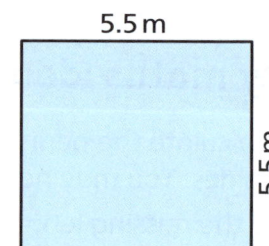

2. Calculate the area of each of these shapes:
   a a square with sides of 12 cm
   b a rectangle that is 12 cm wide and 15 cm long
   c a rectangle that is 6 cm wide and 8.5 cm long
   d a square with sides $3\frac{1}{2}$ cm long
   e a rectangle that is twice as long as it is wide, with a width of 7 cm

3. A rectangular field is 30 m long and 12 m wide. Calculate its perimeter and its area.

Mental maths

1. Look at this growing pattern of squares.
   a What is the perimeter of each square?
   b What pattern can you see in your answer to question **a**?
   c What is the area of each square?
   d How could you describe the pattern in your answer to question **c**?
   e What would the perimeter and area be of the tenth square in this pattern?

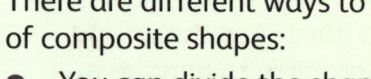
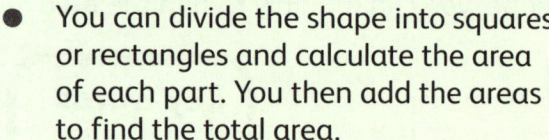
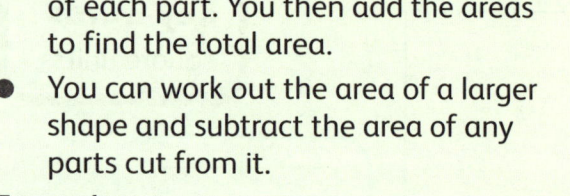

Area of composite shapes

Key maths idea

There are different ways to find the area of composite shapes:
- You can divide the shape into squares or rectangles and calculate the area of each part. You then add the areas to find the total area.
- You can work out the area of a larger shape and subtract the area of any parts cut from it.

Here are two ways to do this.

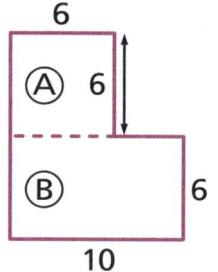

$A = l \times w$
Rectangle A = 6 cm × 6 cm
= 36 cm²
Rectangle B = 10 cm × 6 cm
= 60 cm²
Area A + Area B = 36 cm² + 60 cm²
= 96 cm²

Example
Determine the area of this shape.

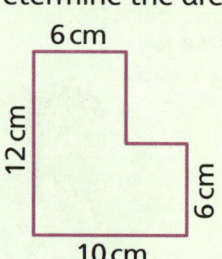

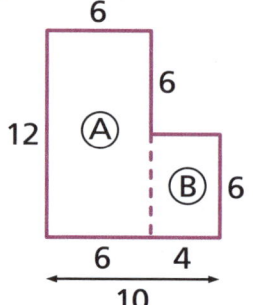

$A = l \times w$
Rectangle A = 12 cm × 6 cm
= 72 cm²
Rectangle B = 4 cm × 6 cm
= 24 cm²
Area A + Area B = 72 cm² + 24 cm²
= 96 cm²

Area of composite shapes

1. Calculate the area of each shape. Draw a rough diagram to show how you divide the shape.

   a 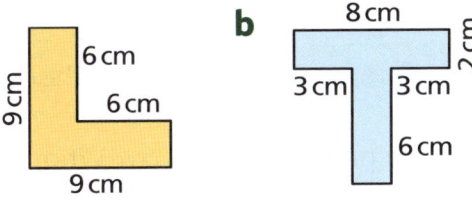 b c 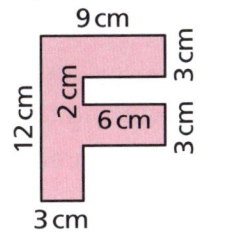 d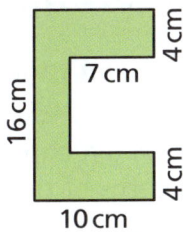

2. Reza and Chin are working on this question. What is the area of this shape?

   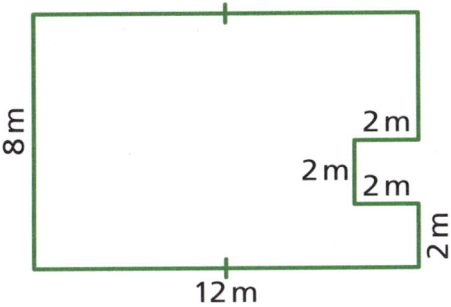

   Reza divides the shape into three rectangles, like this:

   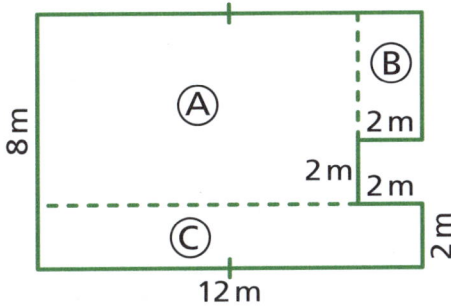

   a Sketch the shape and show two different ways to divide it into rectangles.

   b Explain why you will get the same total area, no matter how you divide up the shape.

   c Chin says you can do fewer calculations if you think of the shape as a rectangle with a square removed, like this:

   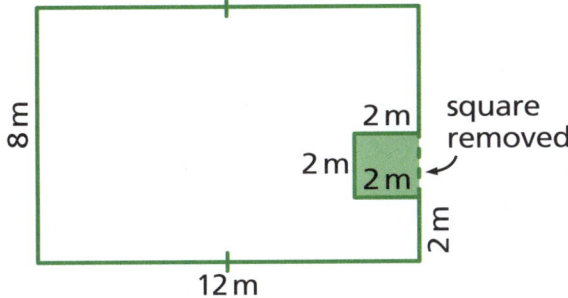

   Work out the area using each person's method. Explain why Chin's method is shorter.

3. Draw the first letter of your name as a composite shape made of rectangles and squares. Calculate the area of the shape.

Full STEAM ahead Design your own picture frames

1. Cut out a rectangle of paper with an area of 30 cm². This will be your artwork. Draw a design of different-coloured quadrilaterals on it.

   When you frame a picture, you normally have a cardboard border around it. This is called a mat.

2. Design two different mats to go around your artwork.
   - The first mat should have an area less than the area of the picture.
   - The second mat should have an area greater than the area of the picture.

   Sketch your designs and label them to show the measurements.

3. Frame costs are worked out by length.
   a Work out the outside perimeter of each mat.
   b Calculate the price of a wooden frame for each mat if the frame costs $2.80 per centimetre.

You will need:
- thick paper or card
- a ruler
- scissors
- coloured markers.

183

Section 11 Measurement Chapter 17 Perimeter and area 2

Problem solving

Choose your own strategy to solve each problem. Estimate before you start and check that your solution is reasonable. Show all the working that you do.

1. Anne has a rectangular cloth with an area of 100 cm^2 and a perimeter of 50 cm. What are the side lengths of the cloth?

2. Raj designs and makes ornaments, like this:
   Each layer is a square.

   a. The top square has a perimeter of 16 cm. What area does it cover?

   b. The bottom square has an area of 100 cm^2. What are the lengths of the sides?

   c. One of the ornaments has a bottom square with sides 12 cm long. If each layer has sides 1 cm shorter than the one below it, and there are eight layers, work out what area of wood Raj would need to build the ornament.

3. Mr Persad has a rectangular lawn that is 5 m long and 4 m wide. He put a paved 1 m-wide path all around the outside of the lawn.

   a. What is the area of the paved path?

   b. He will put an edging strip around the outside edges of the path. What length of edging strip will he need?

   Hint
   How can you use the fact that the perimeter of a square is a multiple of 4 to help you solve this?

4. Randy has two separate square plots of land. The combined perimeter of the two plots is 48 m and the area of the first plot is four times the area of the other. Draw sketches of the two square plots and indicate the lengths of the sides.

5. Zara uses large, rectangular plastic sheets as ground cover during events. She has five different sizes.

Sheet	A	B	C	D	E
Length	6.5 m	11.5 m	6.5 m	0.8 m	3.9 m
Width	3 m	5 m	3.5 m	5.6 m	2.5 m

She wants to cover an area with a total perimeter greater than 36 m but less than 70 m using three sheets. Draw sketches to show two different ways in which she can do this.

What did you learn?

Look back at the work you did in this chapter. Rate your progress.
1 = I cannot do this. **2** = I need more practice. **3** = I understand it and feel confident.

Can you:
- write and explain the formulae for the perimeter and area of squares and rectangles?
- determine the perimeter and area of composite shapes?
- solve problems involving perimeter and area?

Review: Perimeter and area 2

Key terms and concepts

1. Write the words that are missing from these statements about perimeter and area.
   a. ____ is the distance around a shape.
   b. You can use a ____ to calculate the area of a shape if you know the ____ of the sides.
   c. Area is the amount of ____ a flat shape takes up. It is measured in ____ units.
   d. To find the ____ of a rectangle, you can use the formula A = length × width.
   e. To find the area of a composite shape, you can ____ it into smaller squares and rectangles.

Quick check

1. Write down the formula you can use to calculate:
   a. the perimeter of a square
   b. the perimeter of a rectangle
   c. the area of any rectangle
   d. the area of a square.

2. Calculate the area and perimeter of each of these shapes:
   a. a square with sides 3.2 cm long
   b. a rectangle with one side 2.5 m long and the other 3 m long.

3. The perimeter of a rectangle is 20 cm. One side is 8 cm long.
   a. What are the lengths of the other sides?
   b. What is the area of the rectangle?

4. A square has an area of 64 cm². What is the length of one side?

5. A rectangle of area 32 cm² is painted red and blue. The blue painted area is 6.8 cm² smaller than the red area. What area of the rectangle is painted red?

Challenge and investigate

1. Determine the perimeter and area of this field.

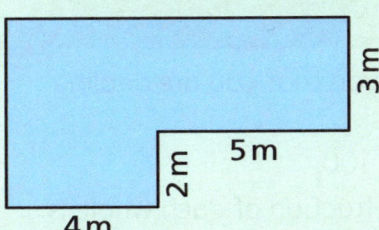

2. Priya has a 40 m² rectangular flower bed that is 8 m long. She wants to make the flower bed bigger, so she makes it 2 m longer. She wants the larger flower bed to have an area of 60 m². How much wider must she make it?

3. Lee wants to paint the wall around his property. One litre of paint can cover 16 m². The wall is 80 m long and 2 m high. How many litres of paint will he need?

4. Jenny buys floor tiles that are 30 cm long and 30 cm wide. How many tiles will she need to tile a rectangular area that is 2.4 m wide and 3.6 m long?

5. Marco cuts a plastic rectangle into two parts. He labels the parts A and B. Area A is $\frac{5}{8}$ of the total area and it is 24 cm² greater than area B. What is the total area of the plastic rectangle.

SECTION 12

Chapter 18 Percentages

In this chapter, you will:
- learn what a percentage is and use the percent (%) symbol
- convert between equivalent fractions, decimals and percentages
- compare and order fractions, decimals and percentages
- calculate percentages of quantities
- express a quantity as a percentage of another quantity
- solve problems involving percentages.

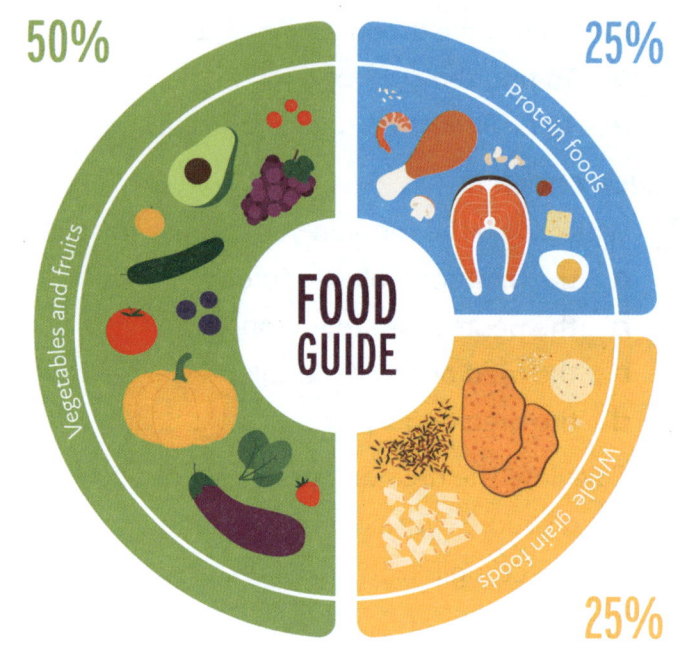

Starting point

Katelyn is doing a project on healthy eating. She found this picture in a magazine.

1. What fractions can you see in the circle?
2. What fraction of the plate is vegetables and fruits?
3. The picture shows that twenty-five **percent** (25%) of your plate should be protein foods.
   a. Explain what this means using fractions.
   b. What does this diagram show you about **percentages**?

Key words
percent
percentage

Key maths idea

Percent means 'out of 100'. The symbol % means percent and it tells you that you are dealing with a percentage.

A percentage is a way of writing a fraction that has a denominator of 100.

Look at these diagrams. There are 100 small squares in each whole. A fraction of each whole is shaded. We can express each fraction as a percentage.

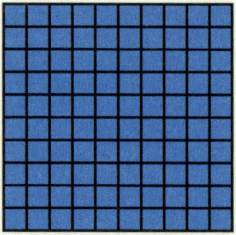

$\frac{100}{100} = 100\%$

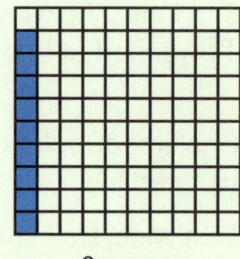

$\frac{9}{100} = 9\%$

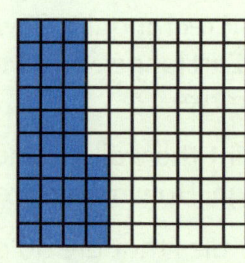

$\frac{34}{100} = 34\%$

186

Equivalent fractions, decimals and percentages

1. For each square, write:
   a the percentage that is shaded
   b the percentage that is not shaded
   c the fraction that is shaded
   d the fraction that is not shaded.

A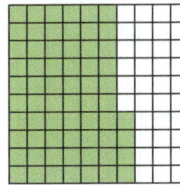

B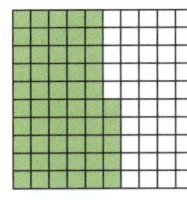

C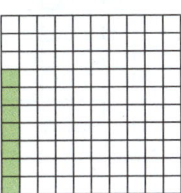

D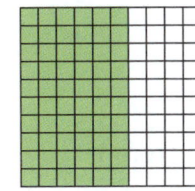

2. Write each percentage as a fraction.
   a 7%   b 11%   c 1%
   d 39%   e 19%   f 83%

3. Write each fraction as a percentage.
   a $\frac{4}{100}$ b $\frac{8}{100}$ c $\frac{29}{100}$ d $\frac{37}{100}$ e $\frac{99}{100}$
   f $\frac{8}{10}$ g $\frac{4}{10}$ h $\frac{1}{10}$ i $\frac{7}{10}$ j $\frac{10}{10}$

Talking maths

1. Look at the diagram.
   a What percentage of your body is water?
   b How could you express that as a fraction?
   c What happens to the amount of water in the body as you get older? How can you tell this from the diagram?

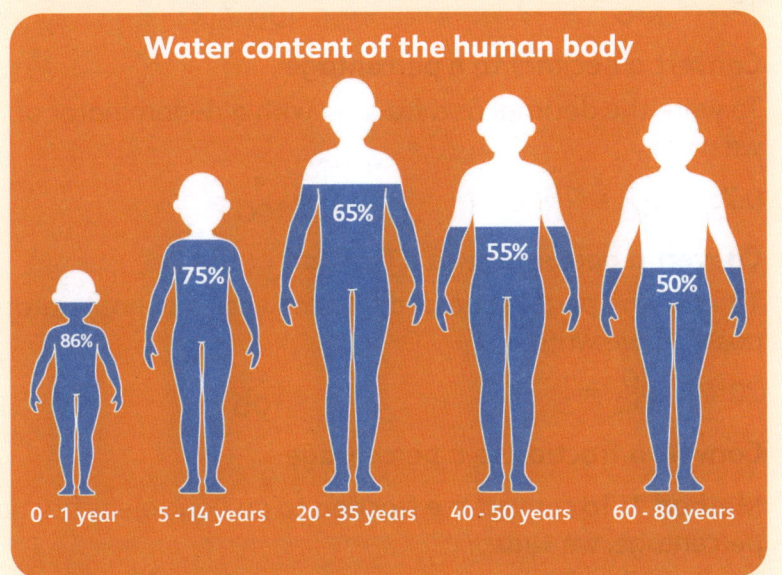

Real-life maths

Percentage symbols can often be found on sale signs in stores. Can you think of anywhere else you might find a percentage symbol?

Section 12 **Number** Chapter 18 Percentages

Equivalent fractions, decimals and percentages

Key maths idea

Look at this shape. 25 out of 100 squares are shaded green. This is one quarter of the whole square. This is also twenty-five hundredths.

We can express the shaded part of the whole as a fraction, as a decimal and as a percentage.

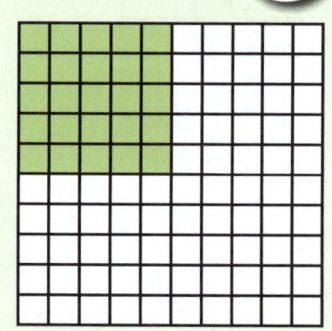

$\frac{25}{100} = 0.25 = 25\%$

$\frac{25}{100}$ can be simplified to $\frac{1}{4}$

So, $\frac{1}{4} = 0.25 = 25\%$

Fractions, decimals and percentages are different ways of describing parts of a whole. You can convert between them, for example you can express a fraction as a decimal, a decimal as a percentage, a percentage as a fraction, and so on.

The number line shows some percentages and their equivalent fractions and decimals.

0%	10%	25%	35%	50%	75%	90%	100%
0	$\frac{1}{10}$	$\frac{1}{4}$	$\frac{7}{20}$	$\frac{1}{2}$	$\frac{3}{4}$	$\frac{9}{10}$	1
0	0.1	0.25	0.35	0.5	0.75	0.9	1

Convert a percentage to a decimal

To write the percentage as a fraction with a denominator of 100, then the fraction as a decimal, we write:

$15\% = \frac{15}{100} = 0.15$ $\qquad$ $3\% = \frac{3}{100} = 0.03$

Convert a decimal to a percentage

To write the decimal as a fraction with a denominator of 100, then the fraction as a percentage, we write:

$0.75 = \frac{75}{100} = 75\%$ $\qquad$ $0.09 = \frac{9}{100} = 9\%$

Convert a percentage to a fraction

To write the percentage as a fraction with a denominator of 100, then simplify the fraction if possible, we write:

$20\% = \frac{20}{100} = \frac{1}{5}$ $\qquad$ $73\% = \frac{73}{100}$ $\qquad$ $8\% = \frac{8}{100} = \frac{2}{25}$

Convert a fraction to a percentage

Method 1: To convert the fraction to an equivalent fraction with a denominator of 100, then to a percentage, we write:

$\frac{3}{4} = \frac{3}{4} \times \frac{25}{25} = \frac{75}{100} = 75\%$ $\qquad$ $\frac{4}{5} = \frac{4}{5} \times \frac{20}{20} = \frac{80}{100} = 80\%$

Method 2: To multiply the fraction by 100% ($\frac{100}{1}$) and simplify to get the percentage, we write:

$\frac{3}{4} \times \frac{100}{1} = \frac{300}{4} = 75\%$ $\qquad$ $\frac{1}{5} \times \frac{100}{1} = \frac{100}{5} = 20\%$

Compare and order fractions, decimals and percentages

1 Write each percentage as an equivalent fraction in its simplest form.
  a 20%   **b** 60%   **c** 80%   **d** 50%   **e** 28%

2 Express the shaded part of each shape as a percentage.
  a   **b**   **c**   **d**
  e   **f**   **g**   **h**

3 Convert each fraction to a percentage.
  a $\frac{30}{100}$ **b** $\frac{8}{10}$ **c** $\frac{8}{25}$ **d** $\frac{9}{50}$ **e** $\frac{4}{5}$
  f $\frac{11}{20}$ **g** $\frac{25}{50}$ **h** $\frac{4}{100}$ **i** $\frac{1}{2}$ **j** $\frac{3}{4}$

Mental maths

1 Convert each decimal to a percentage.
  a 0.45   **b** 0.82   **c** 0.25   **d** 0.66   **e** 0.32
  f 0.9   **g** 0.5   **h** 0.3   **i** 0.08   **j** 0.01

2 Express each percentage as a decimal.
  a 85%   **b** 99%   **c** 35%   **d** 50%   **e** 100%
  f 9%   **g** 3%   **h** 16%   **i** 1%   **j** 120%

Compare and order fractions, decimals and percentages

Key maths idea

Fractions, decimals and percentages can all be written as equivalents of each other. We can compare and order them by converting them into the same form.

Example 1
Which is **greater**: $\frac{7}{25}$ or 25%?
Write $\frac{7}{25}$ as a percentage: $\frac{7}{25} \times \frac{4}{4} = \frac{28}{100} = 28\%$ 28% > 25%, so $\frac{7}{25}$ > 25%

Example 2
Arrange the values 72%, $\frac{19}{25}$ and 0.67 in **ascending** order.

Method 1: Convert all the values to decimals:
72% = 0.72 $\frac{19}{25} = \frac{19}{25} \times \frac{4}{4} = \frac{76}{100} = 0.76$

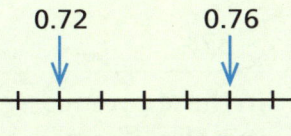

You can show the values on a number line to order them:

Method 2: Convert all the values to percentages: $\frac{19}{25} = \frac{19}{25} \times \frac{4}{4} = \frac{76}{100} = 76\%$ 0.67 × 100 = 67%

Arrange the percentages in ascending order: 67%, 72%, 76%

Always write the answer using the values you were given: **0.67, 72%, $\frac{19}{25}$**

Section 12 Number Chapter 18 Percentages

1 Copy and complete this table to show equivalent forms.

Fraction	$\frac{16}{25}$			$\frac{3}{4}$	
Decimal		0.55			
Percentage			45%		50%

2 Which number is smaller in each pair:
  a 0.9 or 95%? **b** 22% or 0.2? **c** $\frac{31}{50}$ or 31%? **d** 50% or $\frac{11}{20}$?
  e 0.86 or $\frac{44}{50}$? **f** 1.2 or 12%? **g** 75% or $\frac{3}{5}$? **h** 17% or $\frac{4}{25}$?

3 Malaika got 21 out of 25 for a test and Patricia got 83%. Which score is **higher**?

4 Different mammals sleep for different percentages of the day. Order these mammals by the amount of time they sleep.

Animal	Cat	Dog	Koala	Baby
Portion of day spent sleeping	63%	0.42	$\frac{23}{25}$	$\frac{34}{50}$

Think about whether it makes sense to convert them all to fractions, to decimals or to percentages.

Problem solving

1 a Work with a partner. Discuss what strategy you will use to order this set of numbers from **smallest** to **greatest**.

$\frac{4}{5}$ 12% $\frac{7}{20}$ 0.28   0.32   41%   9%   0.85 $\frac{3}{5}$

  b Use your strategy to order the numbers.
  c Draw a number line from 0 to 1 and show each number on it.

2 Write these numbers:
  a a percentage that is $\frac{4}{100}$ less than 0.63
  b a percentage that, when doubled, is equivalent to $\frac{22}{25}$
  c a fraction in its simplest terms equivalent to double the sum of 10% and 0.35.

3 Convert these test scores to percentages and rank them from **highest** to **lowest**.

40 out of 80   90 out of 150   24 out of 30   70 out of 100

4 What whole numbers could you write in place of the letters so that each set of numbers is in **ascending** order?
  a $\frac{1}{A}, \frac{B}{4}, 80\%$ **b** $\frac{C}{5}, 42\%, \frac{24}{D}$ **c** $\frac{2}{E}, 26\%, \frac{F}{3}$

5 Sadia put three numbers in descending order. A common fraction belonging to the eighths family is missing. What fraction could it be?

82%   ☐   0.68

6 In these pyramids, each block is the sum of the two blocks below it. Copy the pyramids and complete them using only percentages.

  a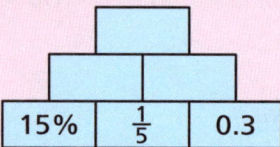

  b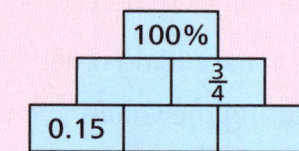

Calculate a percentage of a quantity

Key maths idea

To find a percentage of a quantity, you can write the percentage as a fraction or a decimal and multiply by the quantity.

Example 1
What is 10% of $250?

$10\% = \frac{10}{100} = \frac{1}{10}$

$\frac{1}{10} \times 250 = \frac{1}{10} \times \frac{250}{1}$

$= \frac{250}{10}$

$= \$25$

$10\% = 0.1$

$0.1 \times 250 = 25.0$

$= \$25$

Hint
Remember: the word 'of' tells us to multiply.

Example 2
What is 8% of 120 grams?

$8\% = \frac{8}{100}$

$\frac{8}{100} \times \frac{120}{1}$

$= \frac{960}{100}$

$= 9.6$ grams

$8\% = 0.08$

$0.08 \times 120 = 9.60 = 9.6$ grams

```
  1 2 0
×     8
-------
  9 6 0
```

Hint
Remember: dividing by 100 moves the digits two places to the right.

1 Calculate:
- **a** 2% of 300
- **b** 10% of 500
- **c** 50% of 200
- **d** 20% of 40
- **e** 75% of 80
- **f** 5% of 80
- **g** 100% of 96
- **h** 15% of 60

2 Calculate. Give your answer as a decimal if necessary.
- **a** 30% of 80
- **b** 80% of 260
- **c** 25% of 300
- **d** 60% of 60
- **e** 15% of 300
- **f** 20% of 16
- **g** 5% of 125
- **h** 78% of 144

3 In a school of 400 students, 30% of the students walk to school; the rest use some form of transport.
- **a** How many students walk to school?
- **b** How many students use some form of transport?

4 To reserve a hotel room, Jaden has to pay a 20% deposit before he checks in. If the room costs $178, what deposit will he pay?

Express quantities as percentages

Key maths idea

To write one amount as a percentage of another, you form a fraction and convert it to a percentage.

Example 1
There are 420 students in a school. 126 of the students take part in a maths competition. What percentage of the students take part in the competition?

126 out of 420 students take part

$\frac{126}{420} \div \frac{2}{2} = \frac{63}{210}$ $\frac{63}{210} \div \frac{7}{7} = \frac{9}{30}$ $\frac{9}{30} \div \frac{3}{3} = \frac{3}{10}$

$\frac{3}{10} = 30\%$ So, 30% of the students take part.

Example 2
Sharon has run 17 km of a 20 km road race. What percentage of the race has she completed?

17 km of 20 km $= \frac{17}{20}$ of the race

To convert $\frac{17}{20}$ to a fraction with a denominator of 100, we can multiply by $\frac{5}{5}$

$\frac{17}{20} \times \frac{5}{5} = \frac{85}{100}$

$= 85\%$ She has completed 85% of the race

Section 12 Number Chapter 18 Percentages

1. Calculate:
   a. $8 as a percentage of $40
   b. $9 as a percentage of $25
   c. 12 cents as a percentage of $2
   d. 12 g as a percentage of 300 g
   e. 150 g as a percentage of 1 kg
   f. 70 marks out of 80 as a percentage
   g. 500 ml as a percentage of a litre
   h. 45 minutes as a percentage of 1 hour
   i. 3 months as a percentage of a year
   j. 9 minutes as a percentage of 1 hour

2. Jeevan played 27 games out of 36 games during the season. What percentage of the games did he play?

3. 18 students in a class of 24 voted for Alvin as their class captain. What percentage of the class did not vote for Alvin?

4. In one group, 13 out of 100 students scored an A grade in their mathematics test. In another group, 4 out of every 20 students scored an A grade. What percentage of each group of students scored an A grade?

> **Full STEAM ahead** Collect data to check a prediction
>
> 1. Think about the vowels: a, e, i, o and u. Which vowel do you think is most common?
> 2. Rank the vowels in order from the most common to the least common.
> 3. Choose a passage from your reading book.
> 4. Use tallies to record how many times each vowel occurs in the first 50 words.
> 5. Calculate the percentage use of each vowel.
> 6. Compare the percentages with your ranking. What can you conclude?
>
> **You will need:**
> - a reading book.

Profit, loss and discount

> **Key maths idea**
>
> You can calculate profit or loss as a percentage, if you know the cost price and the selling price.
>
> **Example**
>
> Derrick bought furniture for $12 600. He cleaned it up and sold it for $15 120. Calculate the percentage profit he made.
>
> Selling price − cost price = profit   First work out the profit or loss amount.
>
> $15 120 − $12 600 = $2520
>
> $\frac{2520}{12\,600}$ Make a fraction using the profit amount and the cost price.
>
> Simplify the fraction as much as you can: $\frac{2520}{12\,600} = \frac{252}{1260} \div \frac{4}{4} = \frac{63}{315} \div \frac{3}{3} = \frac{21}{105} \div \frac{3}{3} = \frac{7}{35} \div \frac{7}{7} = \frac{1}{5}$
>
> Convert to a percentage: $\frac{1}{5} \times 100 = \frac{100}{5} = 20\%$ Derrick made 20% profit.
>
> **Hint**
> You learned about profit and loss in Chapter 12.

1. Calculate the amount of profit or loss in dollars.
   a. 25% profit on $300
   b. 10% loss on $640
   c. 15% loss on $200
   d. 25% profit on $280

2. Use the information to work out the percentage profit or loss on each sale.
   a. Cost price $400, sold for $320
   b. Sold for $720, cost $600 to make

Profit, loss and discount

Key maths idea

A **discount** is a reduction in price. For example, during a sale, a store might offer a 20% discount on the price of all items. That means you will pay 20% less than the normal price.

Key word
discount

Example

Mr Warner sells tables for $3500. One of the tables gets slightly scratched in the showroom, so he offers a customer a 15% discount. How much will the customer pay for the table?

Discount: 15% of $3500   Work out the discount amount.

$\frac{15}{100} \times 3500 = 15 \times 35 = \$525$

They will pay $3500 − $525 = $2975   Subtract the discount amount from the price.

1 Look at the advert carefully.
   a What percentage discount is offered during the sale?
   b Calculate the sale price of each pair of shoes.
   c The ladies' shoes do not sell well, so the shop offers an additional 5% discount on those. What is the sale price now?
   d The cost price of a pair of gents' shoes is $82. Calculate the percentage profit or loss when they are sold at the sale price.

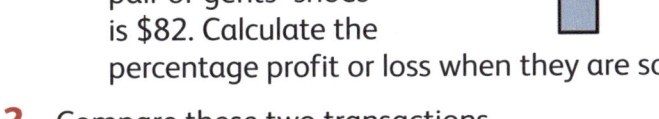

2 Compare these two transactions.

Transaction 1
Cost price: $100
Selling price: $101
Profit: $1

Transaction 2
Cost price: $10
Selling price: $11
Profit: $1

The shopkeeper says she made the same profit ($1) in each transaction.
Discuss the transaction and decide which one was **most** profitable. Explain why.

Real-life maths

Value-added tax (VAT) is a tax that is added to the price of many goods and services. In Trinidad and Tobago, the rate of VAT is 12.5%. Most prices include VAT because the seller has already calculated it and added it to get the price you see.

For example: $80 + 12.5% = $90. You can work out how much VAT to add to a selling price by working out 25% of the price and then halving that amount to get 12.5% (25% of 80 = $20, half of that is $10).

Section 12 **Number** Chapter 18 Percentages

Savings, loans and simple interest

Key maths idea

When you save money, the bank may pay you a percentage of the money you have saved as **interest**.

When you borrow money, you may be charged interest on the **loan** amount.

The amount of money you save or borrow is called the **principal**. The rate of interest paid or charged is a percentage of the principal.

Key words
interest
loan
principal

Example 1

Sasha earns 5% simple interest a year on her savings. How much interest will she earn in a year if she has $10 000 saved?

5% of 10 000 = $\frac{5}{100} \times 10\,000 = \frac{50\,000}{100}$ = $500

Example 2

Reza borrows $800 from Dillon. They agree that Reza will pay 10% simple interest each month till he repays the loan. Reza repays the loan with the interest due after 4 months. How much will he repay in total?

10% of 800 = $80
$80 × 4 = $320
$800 loan + $320 interest = $1120
Reza will repay $1120 in total.

Problem solving

1. Simple interest of $150 was charged on a loan of $750 for one year.
   a. What fraction of the loan is the interest?
   b. What percentage of the loan is the interest?
2. Maleek has to add VAT to the price of furniture that he sells. Calculate the price, including VAT, of the following items.
   a. a desk $360   b. a table $2400   c. a set of chairs $540
3. Ria saves $4000 and earns 8% simple interest per year. How much interest will she earn in two years?
4. Karimah pays 2% simple interest on a loan of $300. Amar pays 2% simple interest on a loan of $450. Explain why they do **not** pay the same amount of interest.

What did you learn?

Look back at the work you did in this chapter. Rate your progress.

1 = I cannot do this. **2** = I need more practice. **3** = I understand it and feel confident.

Can you:
- explain what a percentage is and use the percent (%) symbol?
- convert between equivalent fractions, decimals and percentages?
- compare and order fractions, decimals and percentages?
- calculate percentages of quantities?
- express a quantity as a percentage of another quantity?
- solve problems involving percentages?

194

Review: Percentages

Key terms and concepts

Write short notes to answer these questions and to summarise what you learned in this chapter.

1. How do you order a mixed set of fractions, decimals and percentages?
2. What does 'write one amount as a percentage of another' mean?
3. How do you find a percentage of an amount?
4. What do the terms profit, loss, discount and interest mean?
5. How do you know what you will pay if you are going to get a 20% discount?
6. If you add $25 to a cost price of $100, how do you work out the percentage profit?

Quick check

1. Where do you see fractions, decimals and percentages in everyday life? Make a list.
2. Convert these marks to percentages and write them in **descending** order.

   > 30 out of 50   90 out of 120   48 out of 60   82 out of 100

3. A shopkeeper orders 400 bottles of water. He sells 22% of them on Monday.
   a. What percentage of the bottles are left?
   b. How many bottles did they sell?
   c. If they sell 50% of the remaining bottles on Tuesday, how many bottles will they have left?
4. How much will you save if you buy a plane ticket costing $550 and you get a discount of 20%?

Challenge and investigate

1. a. A part-time worker is paid $30 per hour. She is offered the choice of a 20% increase or an increase of $7.50 per hour. Which option should she take? Why?
   b. A different worker earns $40 per hour. She is given the same choice. Which offer is better for her? Why?
2. A Borough Corporation budgeted $50 000 for refuse removal. At the end of the year, they found they had overspent their budget by 35%. How much did they pay that year for refuse removal?
3. In a school of 900 students, half of them walk to school and 32% travel by bus. The rest travel by other means.
   a. What percentage of the students travel by other means?
   b. Calculate how many students use each form of transport to school and draw a bar chart to show the information. Use a suitable scale for your chart.
4. A computer was sold for $3500. The seller made a 25% profit. What was the cost price of the computer?
5. A store offers 25% discount during a sale. A barbeque set costs $700. What is the sale price of the barbeque?
6. Zaida earns $2000 and Diane earns $2500 dollars. They both get a 10% salary increase. Who receives the **greatest** increase? Why?
7. Safraz buys a laptop costing $3975 and gets a $33\frac{1}{3}$% discount. He also buys a tablet costing $1820 and gets a 20% discount. Calculate how much he saves by buying both items on sale.

SECTION 13

Chapter 19 Handling data 2

In this chapter, you will:
- formulate questions for collecting data
- collect, classify, organise and represent data
- answer questions about data in charts and graphs
- compare different representations of the same data
- suggest sensible decisions based on data
- solve problems involving mode and mean.

Starting point

1. Look at the computer screen. Which of the graphs is:
   a a bar graph?   b a pictogram?   c neither?

2. a Which of the graphs on the computer screen have you seen before?
   b Choose one of the graphs. What questions would you ask to find out what the graph is telling you?

Formulating questions

Key maths idea

Graphs and charts show **data**. Data is another word for information about a group. Before we collect data, we need to **formulate** questions on a topic we want to find out about. A good question should be:
- clear about who we will ask (for example, classmates or family)
- something we can **research**
- interesting and useful to find out about
- fair – it should ask the question clearly, without trying to influence the answer.

Key words
data
formulate
research

Example

The parent committee has some money available for playground equipment. They need to decide what equipment to buy.

They ask the question: Which games are most popular in the school playground during break time?

They can collect data about:
- the number of students playing each game each day in the playground
- the number of students using the existing equipment each day
- what students say they would like to use in the playground.

196

Formulating questions

1 Read each situation. Write a question you could ask that would help you collect useful data.

   a The teachers want to offer three new extra-curricular activities. They want to decide which activities to offer.

   b The school cafeteria wants to offer healthy lunches. They want to decide on three meals to offer.

   c A teacher wants to set up a small class library. He needs to decide what kinds of books to buy.

   Hint
   Make sure you are clear about who you need to ask.

2 These tally charts are missing their headings. Think about the question each tally chart could be answering. Write a suitable heading for each one.

   a
Slippers	卌
Sneakers	卌 II
Sandals	II
? Other	IIII

   b
None	卌
One	卌 IIII
Two	卌 III
More than two	卌

3 For each tally chart in question 2, how many people answered the question?

4 This pictogram shows the weekly pizza sales for three restaurants, but the key is missing. Use this clue to help you work out what the key should be.

   Clue: Luigi's restaurant sold 25 pizzas.

   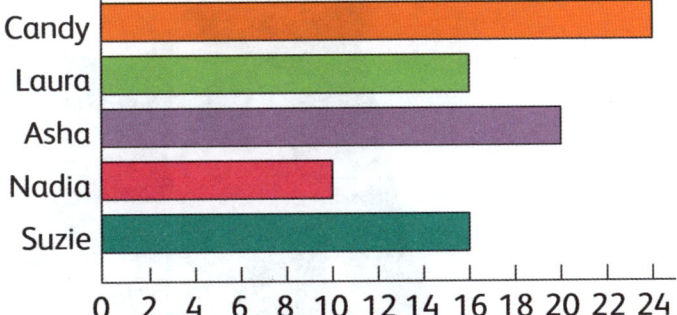

5 This bar chart shows how many books five students read in a year.

   a Write the missing title and axis label.

   Candy — 24
   Laura — 16
   Asha — 20
   Nadia — 10
   Suzie — 16
   0 2 4 6 8 10 12 14 16 18 20 22 24

   b How many books did Candy read?
   c How many more books did Suzie read than Nadia?
   d How many books did Laura and Asha read altogether?

197

Section 13 Statistics Chapter 19 Handling data 2

Collecting, recording and interpreting data

Key maths idea

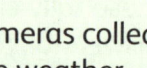

There are many methods to **collect** data.

In an interview, a researcher asks questions. They can **interview** people one at a time, or in groups.

A **survey** collects information from more people. The researcher may send out forms and questionnaires.

Some data is collected by computers, machines or instruments. For example, traffic cameras collect data about the speeds of cars on the road. Weather instruments collect data about the weather.

When you collect data, you can **record** it in different ways. You might write notes, fill in a table or tally chart, or use a questionnaire or form.

We **represent** data in graphs or charts. Then we can see what it means. This is called **interpreting** the data. We can use the data to make decisions.

Key words
collect
interview
survey
record
represent
interpreting
categories

1   Which method would be most effective for collecting data about:
    a   your family?
    b   everyone in your school?
    c   how much rain falls in a year on your island?

2   The pet shop owner collected some data about students who visited the shop. She collected it in a tally chart, then drew this bar chart.
    a   What question do you think this bar chart answers?
    b   What does the horizontal axis show?
    c   What does the vertical axis show?
    d   The data is arranged in five **categories**. What are they?
    e   What is the scale of the horizontal axis?

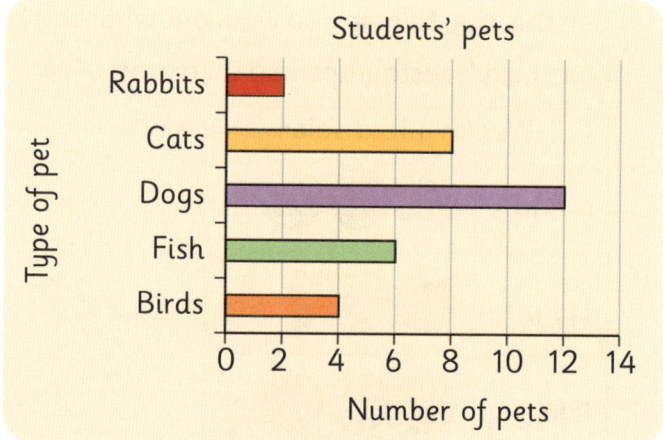

3   The pet shop owner collected the data by talking to children who came into the shop, and by writing down their answers. Which method would be more effective for collecting the data?

4   The pet shop owner says the chart shows she should buy more products for rabbit, fish and bird owners.
    a   Explain her mistake.
    b   Which type of pet food should the owner buy the most of? Explain your answer.

Mode and mean

Problem solving

1. A baker collected this information about the loaves he sold over five days. The baker makes loaves in batches of twelve.
   a. What is the scale of the vertical axis?
   b. How many loaves did the baker sell over the whole week?
   c. He wants to make sure that he sells at least 6 loaves from every batch. How many loaves should he bake on each day? Give reasons.

Mode and mean

Key maths idea

To **analyse** data, look for patterns.

The value that occurs most often is the **mode**. Another useful way to analyse data is to calculate the **mean**. To calculate the mean, follow these steps:

1. Add up all the values.
2. Divide by the number of values.

Key words
analyse
mode
mean

Example

Some students are building towers out of blocks. They can choose between five and ten blocks for each tower. Here is a set of six towers.

6   6   7   6   10   7   7   7   7   7   7   7

The mode is 6. But what is the mean?

Imagine doing this: You put all the blocks from the six towers into one big group. Then you share them out into six equal towers.

6 + 6 + 7 + 6 + 10 + 7 = 42   Add up all the values.
42 ÷ 6 = 7   Divide by the number of values.

Hint
The mean can be a fraction. This happens when the sum of the values is not divisible by the number of values in the set.

The mean is 7.

1. Work out the mode and the mean of each set of values.
   a. 14, 8, 2, 3, 9, 8, 14, 2, 8
   b. 10, 12, 11, 12, 11, 10, 10, 10
   c. 30, 39, 28, 25, 39, 20
   d. 9, 8, 9, 8, 8, 8, 9, 9, 10, 9, 9

2. A data set of five values has a mean of 10. The first four values are 8, 12, 10 and 7. Calculate the missing value.

Hint
Work backwards.

Section 13 Statistics Chapter 19 Handling data 2

3 A data set has seven values. The mean is 15. Here are six of the values: 10, 20, 11, 5, 18, 19. Calculate the missing value.

4 Natasha buys nine 1-kg bags of apples. She writes how many apples she finds in each bag: 6, 7, 9, 7, 6, 6, 7, 10, 6

   a Calculate the mode of the number of apples per bag.

   b Calculate the mean number of apples per bag.

5 What is the mean weight of the apples in a 1-kg bag that contains:

   a 6 apples?   **b** 10 apples?

6 **a** Write the data set of all the students' ages in your class. Work out the mode and mean.

   b What would happen to the mean of the data set if one more student joined the class, and you included their age in the data set?

   c What would happen to the mean if you included a teacher's age in the data set?

   d Why are the two results so different?

> **Full STEAM ahead** Melting ice
>
> Conduct an experiment to time how long it takes an ice cube to melt in the shade and in the sun. Try it with five different ice cubes in each location. Work out the average for each data set.

Analysing data

> **Key maths idea**
>
> We represent data in different ways, such as in tables, graphs or charts. These are called different **representations** of data. They show the data to the **audience** – the people who need to understand and use it. The representation should make the data clear and easy to understand. It also helps us to understand what the data is telling us. This is called **analysing** the data.
>
> **Example**
>
> Malaika did a data-collection project. For her written assignment, she wanted to show the teacher how many students chose each flavour of ice cream. For her presentation, she wanted to show the results in a colourful way that would grab her friends' attention. So, she chose these representations.

Pictograph

Bar chart

Key words
representation
audience
analyse

Analysing data

1. **a** What is unusual about Malaika's bar chart?
   b She forgot to label the bars. Work out which cone represents which flavour.
   c What is the difference between the scale in the pictograph and in the bar chart?

2. A school principal wants to set up a recycling zone at school. She has space for only one type of item: plastic bottles, glass bottles or cans. She decides to run a short experiment to find out what the best choice would be.

   Number of items brought each day

	Plastic bottles	Glass bottles	Cans
Mon	16	5	11
Tue	11	7	10
Wed	7	7	14
Thurs	10	7	13
Fri	11	4	17

   Hint
   What was the same about Monday and Tuesday?

   a On Tuesday, the principal looked at the numbers of each item brought over the first two days. What trend do you think she noticed about these two days that would change by Friday?
   b What is the mode for the data set about glass bottles?
   c What is the mode for the data set about plastic bottles?
   d Copy and complete this frequency chart for the week.

	Frequency
Plastic bottles	
Glass bottles	
Cans	

3. Use the frequency chart from question **2d** to draw a bar chart showing the frequency of each item over the whole week. Follow these steps:
   - Decide whether to make a horizontal or vertical bar chart.
   - Order the bars so that they are arranged from **smallest** to **greatest**.
   - Decide on a suitable scale.
   - Label the axes.

4. Based on the frequency table and bar chart, what decision should the principal make about the recycling station?

5. Which is the most suitable way to show the data to the school to explain the decision: the table, frequency chart or bar chart? Explain your choice.

Talking maths

Keshon says that the colours on a bar chart have no meaning. Why is he correct? Why do you think people use different colours for the bars?

Section 13 **Statistics**   Chapter 19 Handling data 2

Data-collection projects

Key maths idea

The data problem-solving cycle is a long, slow process. These are the main steps:

Identify the problem and formulate the question
- What is an issue or problem you see at school or in your community?
- How could data help solve the problem?
- What question would you ask to collect this data?

Plan
- Where will you collect your data?
- How will you collect your data? Which method would be most effective for your project?
- Who will you ask? How many people?
- Will you ask questions or use data that other people have already collected?
- If you want to do an experiment, what will you need? How will it work?

Collect
- Interview people or do the experiment you planned.
- If you are using data that others have collected, do the research to find the information you need.
- Record your data neatly. You can draw tally charts or frequency tables, or use software such as digital spreadsheets.

Organise and represent
- Put your data in order.
- Choose the best way to represent your data, for example, in charts or graphs.
- Display your data by making a poster or a booklet, or present your work digitally (for example, on a computer).

Analyse and interpret
- Find patterns in your data, such as the mode and the mean.
- Explain what the data tells us about the problem.

Communicate
- Write a summary of the main findings of your data. Present a written or oral report about it.
- Explain each step you followed.
- Explain what the data showed about the problem.
- Suggest some decisions or actions that would make sense based on the data.

Data-collection projects

1. Identify a problem facing students at your school. Use the problem to help you formulate a question.

OR

Choose one of the following projects. Follow the steps provided on page 202 to complete the data problem-solving cycle.

a Carry out a survey on a sample of 50 cars to find out the most popular colour. Use tally marks to record the number of cars per colour (for example, white, black, silver, blue, red, other). Create a table of the results. Then draw a bar graph to represent your data.

b Work in groups. Carry out a 'getting to know you' survey.
- Collect information about the people in your group, for example, how many hours a day they spend on homework, sleeping or playing outside, how many pets they have or how many siblings they have. Make sure that the answers to your questions are numbers or things that you can count.
- Present your data using a graph.
- Use a different colour for each member of the group.
- Each group can then use the graphs to tell the class more about themselves.

Problem solving

1. Five students sold 70 raffle tickets in a week. They started filling in their totals in this table.

Student	Number of tickets sold
Tyrone	12
Abigail	18
Amari	
Javid	14
Gary	12

a How many did Amari sell in the week?
b What was the mean number of tickets the five students sold?
c What was the mode?
d After another week, the mean went up to 18. What was the total number of tickets sold by the end of the second week?

Section 13 Statistics Chapter 19 Handling data 2

(continued)

2. This pictograph uses a scale factor to show how many people were swimming each day at a beach at 12:00 p.m.

Monday	🏊 🏊 🏊 🏊 🏊 🏊
Tuesday	🏊 🏊 🏊 🏊 🏊 🏊 🏊 🏊 🏊
Wednesday	🏊 🏊 🏊
Thursday	🏊 🏊
Friday	🏊 🏊 🏊
Saturday	🏊 🏊 🏊 🏊 🏊 🏊 🏊 🏊 🏊 🏊 🏊 🏊
Sunday	🏊 🏊 🏊 🏊 🏊 🏊 🏊 🏊

a Which day had the most swimmers at 12:00 p.m.?

b Which day had the fewest swimmers?

c On Tuesday, the total number of swimmers was 27. How many more swimmers were in the water on Tuesday than on Monday?

3. This graph is incomplete. It shows how many sandwiches a vendor sold over three days.

The total number of sandwiches she sold altogether in the first two days was 80. How many sandwiches did she sell on Day 3?

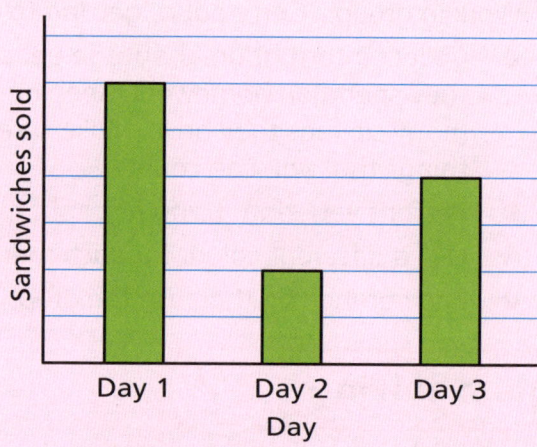

What did you learn?

Look back at the work you did in this chapter. Rate your progress.
1 = I cannot do this. **2** = I need more practice. **3** = I understand it and feel confident.

Can you:
- formulate questions that you can solve using data?
- collect, classify, organise and represent data?
- answer questions about data in charts and graphs?
- compare different representations of the same data?
- suggest sensible decisions based on data?
- solve problems involving mode and mean?

Review: Handling data 2

Key terms and concepts

1 For each definition or example, choose the matching word.

   mode   mean   value   representation   audience

   a The people who look at or listen to a presentation
   b The number that occurs most often
   c Any one of the numbers in a data set
   d A way of showing data, such as a chart or graph
   e To find this, you add all the numbers in a set and divide by how many numbers you added

Quick check

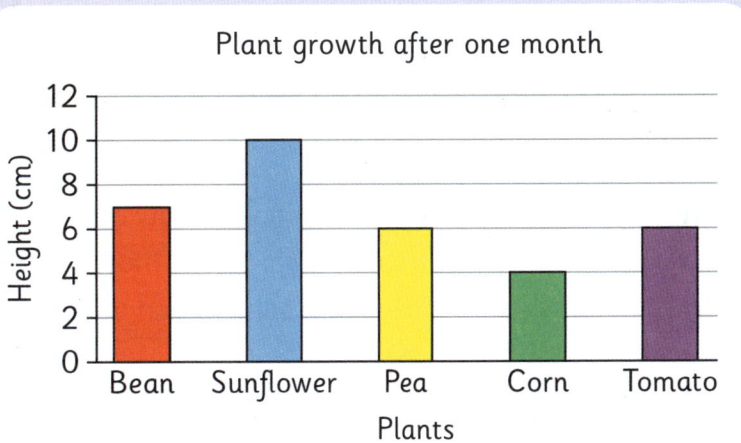

1 This bar chart shows how much five plants grew in one month.

   a Copy and complete the sentences about the chart.

   The vertical axis on this chart shows ____. The horizontal axis shows ____.

2 Describe the scale used on the bar chart.

3 The data in the bar chart came from this tally chart, but some of the tallies have been removed. Write what the tallies should look like for those plants.

1 tally = 1 cm

Bean	
Sunflower	𝍬 𝍬
Pea	
Corn	IIII
Tomato	

4 Work out the mean height of the five plants.

Challenge and investigate

1 Work in pairs or in groups of three. You will need a set of cards numbered from 1 to 20. Mix them up and place them face down on the desk.
   - Take seven cards from the pile.
   - Sort them in order from **smallest** to **greatest**.
   - Find the mean of the set. Round it to the nearest whole number.
   - Is there a mode? If there is, work it out.
   - Compare your results with the other groups. Write the data set of all the means found by all the groups.
   - Identify any trends you notice from your data.

SECTION 14

Chapter 20 Volume

In this chapter, you will:
- use a formula to calculate volumes of cubes and cuboids
- investigate the relationship between length, breadth / width, height and volume
- show that different solids can have the same volume
- solve problems involving volume and capacity.

Starting point

1. **a** Are these boxes cubes or cuboids? How can you tell?
   b Which box do you think takes up the most space? How did you decide?
   c If you filled the boxes with sand, which box would hold the **least** sand? Explain how you decided.
2. **a** Which boxes would hold about the same amount?
   b Is it possible for a taller box to hold the same as a shorter one? Explain why.

Hint
Read through Chapter 11 if you need a reminder of any of these ideas.

Talking maths

Solid objects are three-dimensional.
1. What are the three dimensions of a box-shaped solid?
2. Show your partner where each dimension is on one of the boxes in the picture.

Volume

Key maths idea

Volume is the amount of space that an object occupies. We measure volume in **cubic units**, such as cubic centimetres (cm³) and cubic metres (m³). A cubic centimetre is a cube with its height, length and width all 1 cm.

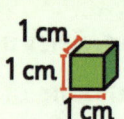

We can put cubes together to make 3-D shapes with different volumes.

This object takes up 5 cubes of space. It has a volume of 5 cubes.

This object has 2 layers of 4 cubes. It has a volume of 8 cubes.

This object has a volume of 15 cubes.

Key words
volume
cubic units

1 Count the number of cubic centimetre units in each solid.

a 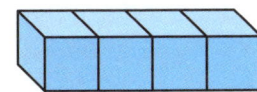   b    c

2 These objects are built out of cubes that each have a volume of 1 cm³. Work out the volume of each object.

a 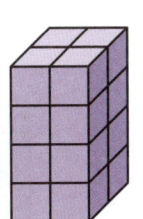   b    c    d

Hint
Remember that 1 cubic centimetre is equivalent to 1 millilitre of water.

3 Jenny has 95 ml of water in a measuring jug. If she dropped shape **a** from question **2** into the water, what would be the new reading on the measuring jug scale? Why?

Finding a formula for volume

Key word
breadth

Key maths idea

To calculate the volume of a cube or cuboid, follow these steps:
1 Work out how many cubes are in one layer.
2 Multiply by the number of layers to find the total.

Look at the cuboid.
One layer has 2 × 4 = 8 cubes.
There are two layers of 8 cubes. 2 × 8 = 16
The cuboid has a volume of 16 cubes.
We can say the cuboid has a length of 4 cubes and a width or **breadth** of 2 cubes.
It has a height of 2 cubes.
The volume of the cuboid is length × breadth × height.
V = l × b × h

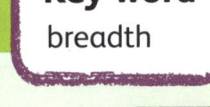

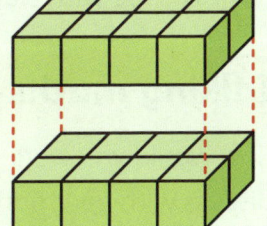
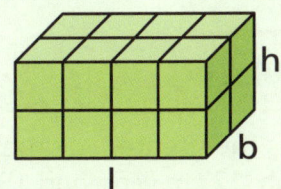

We can use this formula to calculate the volume of cubes and cuboids if we know the length, breadth and height.

Example 1
This brick is 22 cm long, 7 cm high and 10 cm wide.
What is its volume?
Volume = length × breadth × height
= 22 cm × 7 cm × 10 cm
= 1540 cm³

Example 2
What is the volume of a cube with sides 4 cm long?
Volume = length × breadth × height
= 4 cm × 4 cm × 4 cm
= 64 cm³

I know that all the sides of a cube are the same length, so l, b and h are equal.

Section 14 Measurement Chapter 20 Volume

Mental maths

1 Do these multiplications mentally.

a	1 × 1 × 1	b	1 × 2 × 2	c	1 × 1 × 2	d	2 × 2 × 2
e	2 × 2 × 10	f	1 × 2 × 3	g	2 × 2 × 3	h	3 × 3 × 2
i	2 × 4 × 2	j	2 × 5 × 2	k	1 × 3 × 9	l	3 × 3 × 3

1 Apply the formula V = l × b × h to calculate the volume of each stack of cubes.

a 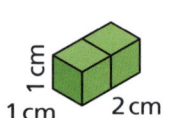 b 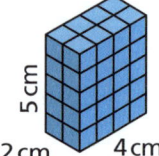 c 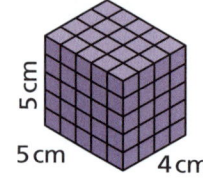 d 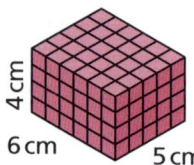 e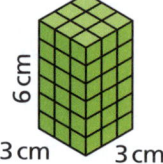

2 Apply the formula to calculate the area of each solid.
  a A cube with sides of 5 cm
  b A cuboid that is 2 cm wide, 3 cm long and 1 cm high
  c A cuboid that is 3 cm wide, 2 cm high and 4 cm long
  d A cube with sides of 6 m
  e A cuboid that is 12 mm long, half as wide as it is long and one third as tall as it is long
  f A cube with sides that are $1\frac{1}{2}$ cm long

Talking maths

How would you work out the volume of this maths book?
Discuss with your partner what you would need to do.

Problem solving

1 Which has the **greater** volume: a brick 12 cm wide, 8 cm high and 23 cm long or a sponge 13 cm high, 12 cm wide and 15 cm long?

2 The volume of a cube is 2 cm³. How many of those cubes could fit into a box that is 10 cm long, 4 cm wide and 6 cm high?

Hint
Draw a diagram of the cube and label the sides in centimetres.

3 A cube with sides 1 cm long has a volume of 1 cm³. What would be the volume of the cube in mm³?

4 How many millilitres of liquid will it take to fill a rectangular container that is 10 cm long, 7 cm wide and 3 cm high?

5 Sadia has 480 ml of water in a 1-litre measuring jug. A cube with sides 5 cm long is placed into the jug and it sinks to the bottom. What would you expect the reading on the scale of the jug to be after the cube is dropped into it?

6 There is 275 ml of water in a jug. When a carrot is placed in the jug the water level rises to 450 ml. What is the volume of the carrot in cm³?

Solving problems involving volume

Full STEAM ahead — Same volume, different shape

Activity 1

There are only two possible solids that you can build by joining three cubes with their faces touching.
What can you say about the volume of each solid?

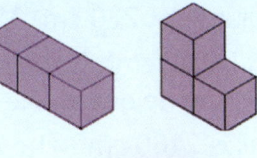

Activity 2

You are going to find eight different ways of forming a solid by joining 4 cubes with their faces touching.

a Stack the cubes to find the eight different ways of making a solid.
b Sketch or take photos of the solids you build.
c How many of the shapes are cuboids?
d Write the calculation to find the volume of the cuboids. What do you notice?

You will need:
- four cubes with the same volume, for example, dice, ones cubes, wooden blocks or cube-shaped beads.

Solving problems involving volume

Key maths idea

You have learned that volume = length × breadth × height.
You can find the length, breadth or height of a cuboid if you know the volume and the other two dimensions.

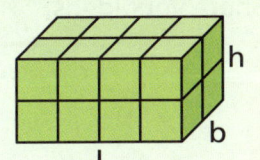

Hint
Remember: breadth means the same as width.

Example 1

A cuboid has a volume of 24 cm³.
It is 3 cm wide and 4 cm long. How high is it?
Start by drawing a sketch to show the information you have.

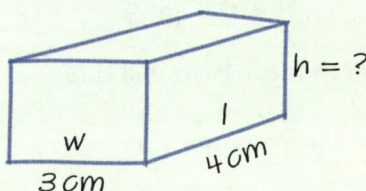

V = l × b × h
24 = 3 × 4 × h
24 = 12 × h
So h must be 2 cm.
Check: 3 × 4 × 2 = 12 × 2 = 24
The cuboid is 2 cm high.

Example 2

A cube has a volume of 27 cm³.
What are the lengths of its sides?
Draw a sketch to visualise the problem.

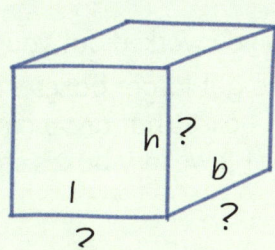

27 = l × b × h
But: l = b = h in a cube, so 27 = l × l × l
First make a guess but then check and improve your strategy to find the answer:
2 × 2 × 2 = 4 × 2 = 8 ✗ too low
3 × 3 × 3 = 9 × 3 = 27 ✓
The sides are 3 cm long.

Section 14 **Measurement** Chapter 20 Volume

Problem solving

1. This cuboid has a volume of 220 cm³. What is its length?
2. The volume of a box is 162 cm³. If the length is 9 cm and the width is 6 cm, what is the height of the box?
3. The volume of a tank is 2160 cm³. The tank is 15 cm long, and its width and height are the same as each other. Sketch the tank and show its dimensions.

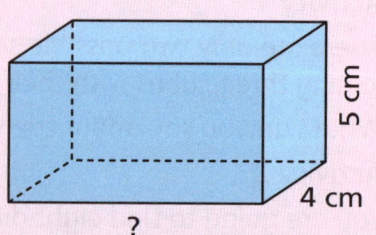

Real-life maths

Airlines have restrictions on the size of suitcases that you may travel with. These are normally given in the form of height × width × length, and they may include the mass. For example, a carry-on bag cannot be larger than 23 cm × 36 cm × 56 cm.

When you buy a suitcase, the capacity is often given in litres.

Talking maths

A carry-on bag is advertised as having a capacity of 30 litres. How do you think the manufacturer works out this figure? Share your ideas.

Problem solving

1. Mariah sells small cube-shaped jewellery boxes that have a volume of 8 cm³. She wants to pack these into a larger box, like the one in the picture.
   a. Investigate how many jewellery boxes can fit into the larger box.
   b. Tell your partner how you worked out your answer.
   c. When Mariah goes to get larger boxes, she finds the supplier only has boxes that are 6 cm wide, 12 cm long and 4 cm high. How will this change the number of jewellery boxes she can pack into one box?

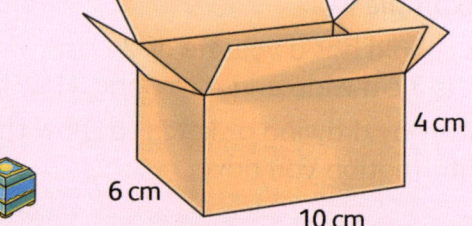

What did you learn?

Look back at the work you did in this chapter. Rate your progress.
1 = I cannot do this. 2 = I need more practice. 3 = I understand it and feel confident.

Can you:
- calculate the number of cubes in a solid?
- use a formula to calculate the volume of a cube or cuboid?
- find one dimension of a solid when the volume and other dimensions are given?
- solve problems involving volume and capacity?

Review: Volume

Key terms and concepts

1 Complete the sentences to summarise what you learned in this chapter.
  a Cubes and cuboids have three dimensions. These are: ____.
  b Volume is a measure of how much ____.
  c The formula for calculating the volume of a cuboid is ____.

Quick check

1 Calculate the volume of each solid.

  a 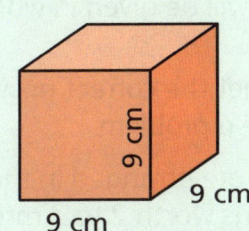 **b** 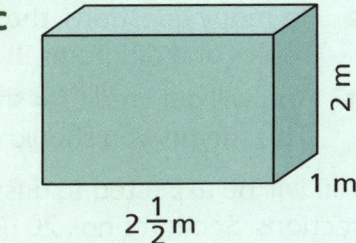 **c**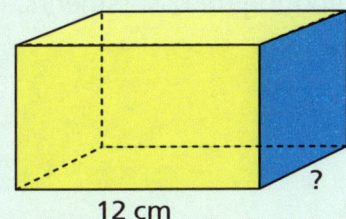

2 Use the formula $V = l \times b \times h$ to calculate the volume of:
  a a cube with sides 5 cm long
  b a cuboid 3 cm high, 5 cm wide and 10 cm long
  c a cuboid that is 3 cm long, 2 cm wide and $\frac{1}{2}$ cm high.

3 The dimensions of a cuboid are 1 cm × 2 cm × 2 cm.
  a If the dimensions were all doubled, what would the new volume be?
  b How many of the original cuboid would fit into the larger one?

Challenge and investigate

1 Use your own shoe for this activity.
  a Measure the length, height and width of your shoe, correct to the nearest centimetre.
  b What is the smallest size box that your shoe could fit into? Sketch the box and label it to show the dimensions.
  c Explain how you worked out the size of the box.

2 Selina is going to make wooden boxes. Each box must have a volume of 2000 cm³ and one of the dimensions (length, width or height) must be 25 cm.
  a What are some possible dimensions for the boxes? Find as many options as you can.
  b Which of the dimensions you found would be practical for wooden boxes? Why?
  c Which dimensions would be impractical? Why?
  d Selina asks you which dimensions you think are most suitable for the boxes. What would you tell her? Give a reason for your answer.

3 The blue shaded face of this cuboid is a square.
  a The volume of the cuboid is 192 cm³. What is its height?
  b If the container was one third full of water, how many millilitres of water would you need to add to fill it?

SECTION 15

Chapter 21 Prepare for the Secondary Entrance Assessment (SEA)

On completion of Standard 5, every student in the country completes the same SEA. The information in this section and the practice questions that follow will help you prepare well for this assessment. It will make sure that you know what to expect and how to answer questions so that you can do your very best.

Just like in maths classes …

- You will have to answer questions from different topics.
- In many questions, the information that you need to answer the question will be given in words, tables and diagrams. It is important to read these carefully.
- You will get credit for showing your ideas and working, even if you do not get the correct answers. This means you should always write each step of your working as you solve a problem.

You will be expected to answer 40 questions totalling 75 marks. The questions are arranged in three sections. Section I has 20 questions worth 1 mark each. Section II has questions worth 2 to 3 marks each and Section III has questions worth 4 marks each. You will have 75 minutes (1 hour and 15 minutes) to complete the test.

Preparing for the SEA

If you want to be good at a sport, you need to stay fit, practise your moves and develop your skills. This idea applies to maths as well. Practise and develop your maths skills so you are ready for the SEA exam.

There are different SEA-type questions in this section. You can use them to prepare for your SEA. The flow diagram shows how you can revise by doing SEA-type questions.

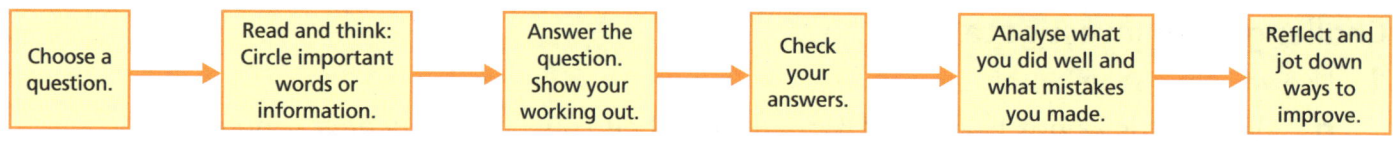

When you are preparing for the SEA, try to do a few questions each day. This will help you develop good techniques. Spending 15 or 20 minutes practising each day is more effective than spending 2 hours once a week.

Learning how to answer SEA-type questions can help you feel less anxious and pressured. It will help you know you are properly prepared and have practised lots of different questions on the topics you have learned.

Improve your well-being

Stress and pressure can affect your performance. If you are tired, you might make careless mistakes. It is important that you take good care of yourself. Make sure you eat well and exercise while you are preparing for assessments.

Read these thoughts. Do you ever think like this?

I am really scared that I will not do well on this task.

I feel like everyone else is going to do well and that I am not.

I hate doing tests; I always freeze up.

I just hope I can work fast enough to answer all the questions.

What am I going to do if I do not understand the question?

Everyone expects me to do brilliantly on the test. What if I do not?

I rushed this morning and now I feel all flustered.

When you feel anxious or stressed out, you can try a technique called box-breathing. Fire-fighters and other emergency workers use this technique to help them remain calm in difficult situations.

Step 1: Breathe in, counting to four slowly. Feel the air enter your lungs.

Step 2: Hold your breath for 4 seconds. Try to avoid inhaling or exhaling for 4 seconds.

Step 3: Slowly exhale through your mouth for 4 seconds.

Step 4: Wait 4 seconds before inhaling again.

Repeat the steps till you feel calm.

Section 15 Exam preparation Chapter 21 Prepare for the Secondary Entrance Assessment (SEA)

Answering SEA questions

Different questions need different types of answers

When you read through the exam paper, you will notice that some questions are simple and others are more complex.

The following examples give you some guidelines on how to answer different types of questions.

> As you work through the questions, think about how you would answer each one. There is usually more than one way to answer a question.

1 mark questions

If a question is worth 1 mark, you can do a single calculation to work it out or just write the answer.

Question 1 (1 mark)

Round 6467 to the nearest thousand.

Answer: **6000** ✓

Question 2 (1 mark)

How many $\frac{1}{4}$ litre bottles can you fill if you have $1\frac{1}{2}$ litres of water?

$1 = \frac{4}{4}$ and $\frac{1}{2} = \frac{2}{4}$

Answer: **6 ($\frac{1}{4}$ litre bottles)** ✓

Question 3 (1 mark)

Terique started his homework at the time shown on the clock.

He finished 45 minutes later. What time did he finish?

Answer: **25 to 4** ✓

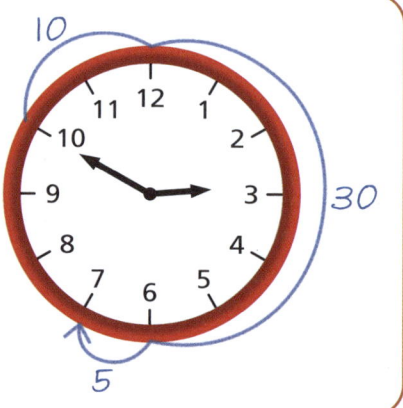

> **Hint**
> Remember: you can draw on the diagrams to help you find the answers.
> You could also write the time as 3.35 or 15:35.

Question 4 (1 mark)

The table shows the number of hours that Mona spent studying each week for a month. Complete the table to show how many hours she spent studying in Week 3 if she spent 30 hours studying that month.

Time spent studying

Week	Time spent (hours)
1	6
2	3
3	**9** ✓
4	12

$6 + 3 + 12 = 21$
$30 - 21 = 9$

> **Hint**
> You can write notes to work out the answers if you need to.

214

Answering SEA questions

2 and 3 mark questions

If a question is worth 2 or 3 marks, you will get marks for showing your working, even if you get the answer incorrect.

Question 1 (2 marks)

Which two square numbers sum to 100?

Square numbers: 1, 4, 9, 16, 25, <u>36</u>, 49, <u>64</u>, 81 ... ✓

36 + 64 = 100 ✓

Question 2 (2 marks)

Katelyn needs 2 m of ribbon for a project. She has 153 cm. How much more ribbon does she need?

2 m = 200 cm

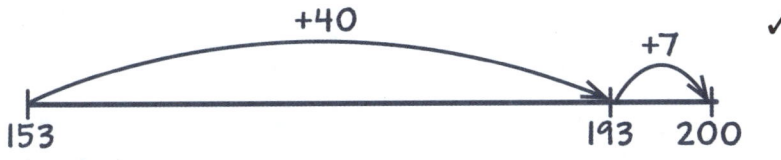

Answer: 47 cm more ✓

Question 3 (3 marks)

Billy shared 85 mangos among himself and his friends Nigel and Dinesh. Billy ended up with 10 more than Nigel, while Dinesh ended up with 20 more than Billy.

How many mangoes did each person end up with?

B	N	D
? + 10	?	? + 10 + 20

85 − 40 = 45

45 ÷ 3 = 15

15 + 10 = 25	15	15 + 30 = 45 ✓

Check: 25 + 15 + 45 = 85 ✓

Answer:

Billy: 25   Nigel: 15   Dinesh: 45 ✓

Question 4 (3 marks)

A computer store has a sale on laptops and tablets. What is the total discount given to a customer who buys these two items?

$33\frac{1}{3}\% = \frac{1}{3}$

$\frac{1}{3}$ of 3360 = 1120 ✓

$25\% = \frac{1}{4}$

$\frac{1}{4}$ of 1600 = 400 ✓

Total discount is $1120 + $400 = $1520 ✓

Answer: $1520

Section 15 Exam preparation   Chapter 21 Prepare for the Secondary Entrance Assessment (SEA)

Question 5 (3 marks)

Show how you could share three whole pizzas among 24 students so that each person gets an equal share. Express each student's share as a fraction of a pizza.

3 pizzas divided among 24 ✓

Hint
This is a good place to use diagrams to help you work out the answer.

Pizza 1   Pizza 2   Pizza 3

$\frac{3 \text{ pizzas}}{24 \text{ students}}$ $\frac{3}{24} = \frac{1}{8}$

$24 \div 3 = 8$ We need 8 slices from each pizza.

Each student will get 1 slice, which is $\frac{1}{8}$ of a pizza. ✓ ✓

4 mark questions

If a question is worth 4 marks, you are expected to show your working and to make your reasoning clear. You can do drawings, write answers or do calculations to show the examiner how you found the solution. You will get marks for correct steps in your working, even if you get the incorrect answer.

Question 1 (4 marks)

The diagram represents a plot of land that is planted with sweet potato and cassava crops. The smaller area is planted with sweet potatoes. What is the area of the land planted with cassava? Show all your working.

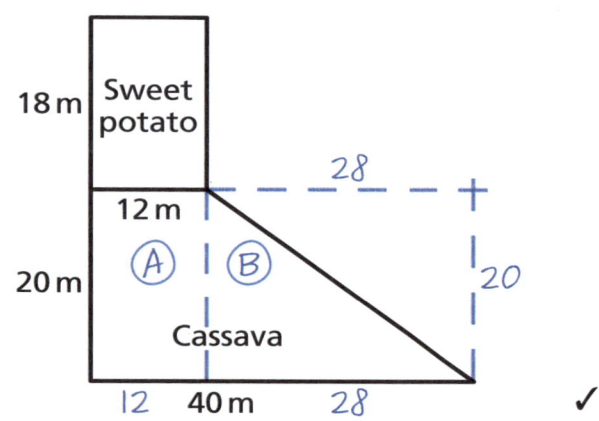

Area A = 20 × 12 m
      = 240 m² ✓

Area B is half a rectangle with area 20 × 28
20 × 28 = 560
$\frac{1}{2}$ of 560 = 280 m² ✓

Area A + Area B = 240 + 280
              = 520 m² ✓

Answer: **520 m²**

Now that you have seen how to answer different kinds of questions, it is time for you to practise on your own.

SEA practice questions

Section I

The following questions are of the types you can expect in Section I. They are worth 1 mark each and you may write the answer only. You may show your working. Working through these examples will help you prepare well for this part of the assessment.

1 Round 1058 to the nearest hundred.
2 Express 0.05 as a percentage.
3 If 25% of a class of 36 students are absent, how many students are present?
4 Express 82% as a decimal.
5 Express 500 minutes in hours.
6 A white boundary line is to be painted around a rectangular area. If the area is 36 km long and 12 km wide, how long will the white line be?
7 The area of a square is 400 cm². What is its perimeter?
8 What is the value of the digit 4 in the number 12.34?
9 This clock is 20 minutes slow. What is the correct time?

10 Vanessa saved half her money. If she saved $12.25, how much money did she have?
11 Patricia picked 24 mangoes. She ate $\frac{1}{8}$ of them. How many did she eat?
12 Jabari planted 6 rows of sweet potato plants, with 12 plants in each row. How many plants is that in total?
13 A rectangle has one side 4 cm long and another 9 cm long. What is the area of the rectangle?
14 How many weeks are there in 84 days?
15 What is $\frac{3}{8} + \frac{2}{8}$?
16 What is $\frac{3}{4}$ of 40?
17 Write the numeral to represent four hundred and three thousand and fifty-seven.
18 The table shows the number of views that an online video received on each of four days. On which day did the video receive the most views?
19 Diane has these items on display on her stall at the beach. What fraction of the items are sunglasses?

Hint
Draw a sketch if you need to.

Day	Number of views
Monday	53 867
Tuesday	53 768
Wednesday	52 876
Thursday	53 876

Hint
Always reduce fractions to their simplest form.

Section 15 Exam preparation Chapter 21 Prepare for the Secondary Entrance Assessment (SEA)

20 One sixth of Tricia's weekly spending on transport is $20. Calculate her weekly spend.

21 Match each of these numbers to its approximate position on the number line: 3.95, 2.95, 3.21

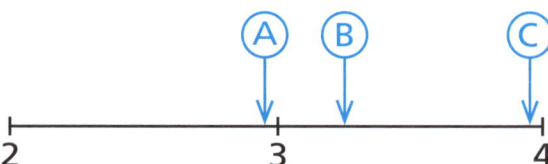

22 Express the yellow shaded area as a decimal that represents part of the whole.

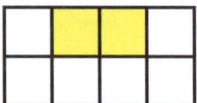

23 Raj has 7 bills totalling thirty-eight dollars. Here are the values of 5 bills:

| $20 | $5 | $5 | $1 | $1 | ? | ? |

What are the values of the other two bills?

24 Ashanti buys three boxes of cupcakes. She makes packages of cupcakes using $\frac{3}{4}$ of each box. How many packages can she make using all three boxes of cupcakes?

25 Convert $3\frac{2}{5}$ hours to minutes.

26 Marcus made this model using 1 cm cubes. What is the volume of the model?

27 Write the correct mathematical name for the quadrilateral below that has no right angles.

28 What is the mode of the following set of test results?

24	25	24	26	28	30
24	30	24	28	27	23
25	24	27	22	24	25

SEA practice questions

29 The graph on the right shows the results of a survey at school. How many more students chose basketball than soccer?

30 If ☐ × 0.04 = 400, what value should go in the box?

31 Which of these letters is symmetrical?
   P   S   B   L

32 What is the difference between the values of the digit 4 in the number 4040?

33 The area of a rectangle is 42 cm². If one side of the shape is 7 cm long, what are the lengths of the other three sides?

34 Omar uses $\frac{3}{4}$ kg of flour to make 15 roti. How many kilograms of flour will he need to make 30 roti?

35 Write 24 as a product of its prime factors.

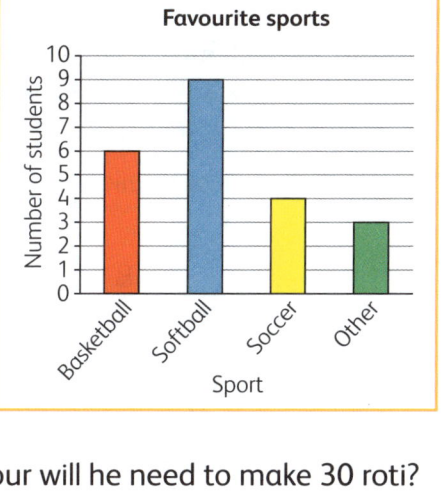

Section II

The questions that follow are of the types you can expect in Section II. These questions are all worth 2 or 3 marks. You will get some marks for showing your working, even if you do not get the correct answer. Practising questions like this will prepare you well for this part of the assessment.

1 Maggie had $39.80. She spent $24.60 on a book, $3.04 on a snack and 98 cents on a small pack of mints. How much money does she have left over?

2 What is the area of the shaded portion of this shape?

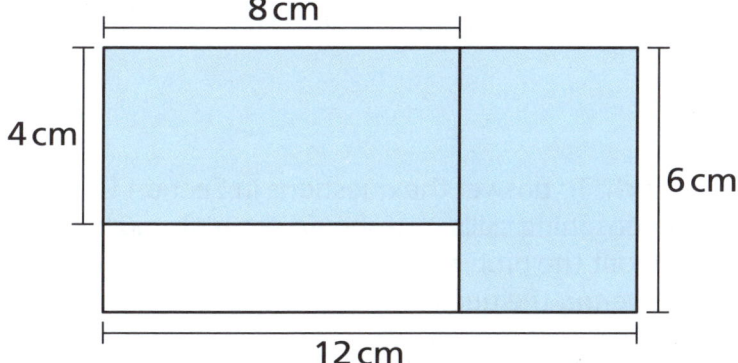

Not to scale

3 A 60 metre length of rope is cut into two pieces. One piece is 3 times as long as the other. What are the lengths of each piece?

4 A pencil is 15 cm long. The pencil is reduced by $\frac{1}{5}$ of its length each time it is sharpened. What will the length be in centimetres after it is sharpened for the second time?

5 Oneika has $22.50 and Jenny has $43.75. If Oneika gets another $9.75 and Jenny spends $12.50, how much money will they have in total?

6 These are the ingredients for mango ice cream:
   400 ml cream
   200 ml milk
   320 g mango
   100 g sugar
   25 ml lemon juice

   If Jemila only has 300 ml of cream, how much mango should she use for this recipe?

Section 15 Exam preparation Chapter 21 Prepare for the Secondary Entrance Assessment (SEA)

7 Randy and Priya share $380. Priya gets $40 more than Randy. How much do they each get?

8 Shop A and Shop B sell the same packets of fruit punch. Zara pays $30 for four packets at Shop A. Keisha pays $30 for three packets and a box of cookies at Shop B. The box of cookies costs $4.20. What is the difference in the price of a packet of fruit punch at Shop A and Shop B?

9 Three identical equilateral triangles are placed next to each other, like this, to form the trapezium ABCD.

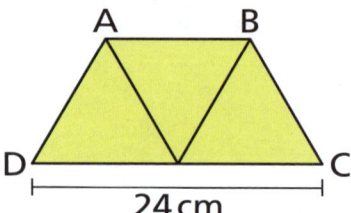

Calculate the difference between the perimeter of one triangle and the perimeter of the trapezium ABCD.

10 An airport clock in Trinidad shows the time as 14:00. The clock is 5 minutes slow. The time in London is 5 hours ahead of the time in Trinidad. What is the correct time in London?

11 Sketch and name a quadrilateral with two lines of symmetry, two pairs of parallel sides and no right angles.

12 The incomplete table shows Mariah's marks as percentages.

Subject	Science	Art	Mathematics	Spanish
Percentage	73	76		83

Mariah's mathematics mark was 42 out of 60. Copy and complete the table and calculate her mean mark for these four subjects.

Section III

The questions in Section III are all worth 4 marks each. To answer the questions in Section III, you need to apply what you have learned and use your reasoning skills. You should show all your working so that the examiner can see how you have thought about the problem and what you have done to solve it. Practising will help you develop strategies for answering the questions in this section.

1 A school places large wooden cubes together to make benches for a school play, like this:

   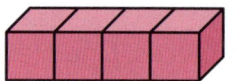

All visible faces will be painted. The face on the ground and the faces that touch each other will not be painted.

Number of cubes in a row (c)	1	2	3	4	5	6
Number of faces to be painted (n)	5					

   a Copy and complete the table to show how many faces will be painted for different numbers of cubes.

   b Write a rule that will allow the school to work out how many faces it needs to paint for any number of blocks.

   c Use your rule to find the number of faces that will need to be painted if there are 10 cubes in the row.

SEA practice questions

2 Josiah and Chin played the same computer game 5 times. Each player's scores are shown in the table.

Josiah	1220	1108	989	1134	1065
Chin	232	808	912	1160	1345

   a Calculate each player's mean score.
   b Who do you think is the best player? Explain your answer.

3 It takes a builder three weeks to build a house. It takes $\frac{2}{3}$ of a week to construct the frame of the house and $1\frac{1}{4}$ weeks to complete the roof. The builder spends the rest of her time on other tasks.
   a What fraction of her time does she spend on other tasks?
   b The builder works 6 days a week. How many days does she spend on other tasks?

4 A rectangular pool 3 m wide and 5 m long is surrounded by a path 1 m wide.
   a What is the area of the path?
   b If the pool is 2 m deep, what volume of water can it hold?

5 The red lines on the diagram are lines of symmetry. Complete the figure so that it is symmetrical and determine its area in square metres.

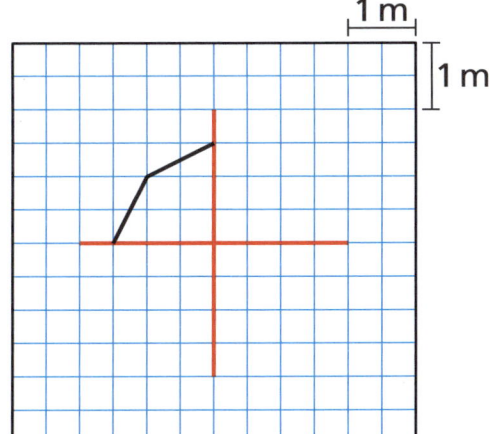

6 Kimani works in a shop 4.3 hours a day. He is paid $28.50 per hour.
   a Calculate how much he is paid per week if he works 5 days a week.
   b One of Kimani's jobs is to make 15 smaller bags of candies from a larger bag weighing 2.4 kg. How many grams should he put into each of the smaller bags?

7 A nursery has 500 trees on sale. On Monday they sold half of the trees. Each day after that, they sold 20% of the remaining trees.

How many trees were left at the end of the work day on Thursday?

8 The sketch shows a shape made by combining a square and a right-angled triangle. The total area of the shape is 40 cm².

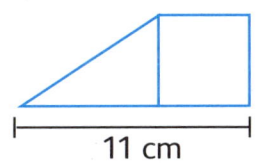

What could the dimensions of the square be? Show how you work this out.